Advances in
MICROBIAL ECOLOGY

Volume 7

ADVANCES IN MICROBIAL ECOLOGY

A Continuation Order Plan is available for this series. A continuation order will bring delivery of
each new volume immediately upon publication. Volumes are billed only upon actual shipment.
For further information please contact the publisher.

Advances in
MICROBIAL ECOLOGY

Volume 7

Edited by

K. C. Marshall

University of New South Wales
Kensington, New South Wales, Australia

PLENUM PRESS · NEW YORK AND LONDON

The Library of Congress cataloged the first volume of this title as follows:

Advances in microbial ecology. v. 1–
 New York, Plenum Press c1977–
 v. ill. 24 cm.
 Key title: Advances in microbial ecology, ISSN 0147-4863
 1. Microbial ecology — Collected works.
QR100.A36   576′.15   77-649698

Library of Congress Catalog Card Number 77-649698
ISBN 0-306-41558-9

© 1984 Plenum Press, New York
A Division of Plenum Publishing Corporation
233 Spring Street, New York, N.Y. 10013

Printed in the United States of America

Contributors

Ronald M. Atlas, Department of Biology, University of Louisville, Louisville, Kentucky 40292

A. W. Bourquin, Environmental Research Laboratory, U.S. Environmental Protection Agency, Gulf Breeze, Florida 32561

David B. Nedwell, Department of Biology, University of Essex, Colchester C04 3SQ, Essex, United Kingdom

P. H. Pritchard, Environmental Research Laboratory, U.S. Environmental Protection Agency, Gulf Breeze, Florida 32561

J. Skujiņš, Department of Biology and Ecology Center, Utah State University, Logan, Utah 84322

Preface

Since the appearance of the first volume of *Advances in Microbial Ecology* in 1977 under the editorship of Martin Alexander, the series has achieved wide recognition as a source of in-depth, critical, and sometimes provocative reviews on the ecology of microorganisms in natural and man-made ecosystems. Most reviews published in *Advances* have been prepared by experts at the invitation of the Editorial Board. The Board intends to continue its policy of soliciting reviews, but individuals are encouraged to submit outlines of unsolicited contributions for consideration of their suitability for publication in *Advances.*

Volume 7 of *Advances in Microbial Ecology* covers a range of topics related to the ecology of microorganisms in natural and artificial habitats. R. M. Atlas discusses the measurement and significance of diversity in microbial communities. The nature of deserts and the activity of microorganisms in desert soils are considered by J. Skujiņš. D. B. Nedwell examines both the input and the mineralization of organic carbon in anaerobic aquatic sediments. The role of microcosms in the evaluation of interactions between pollutants and microorganisms is the basis of a major review by P. H. Pritchard and A. W. Bourquin.

The Editor and members of the Editorial Board of *Advances in Microbial Ecology* are appointed by the International Committee on Microbial Ecology (ICOME) for fixed terms. Martin Alexander and Thomas Rosswall have completed their terms as Board members, and we wish to offer them sincere thanks for their efforts in establishing the series. With the publication of this volume, we welcome Ron Atlas and Bo Barker Jørgensen to the Editorial Board.

K. C. Marshall, Editor
R. Atlas
B. B. Jørgensen
J. H. Slater

Contents

Chapter 3

The Input and Mineralization of Organic Carbon in Anaerobic Aquatic Sediments

David B. Nedwell

Chapter 4

The Use of Microcosms for Evaluation of Interactions between Pollutants and Microorganisms

P. H. Pritchard and A. W. Bourquin

1

Diversity of Microbial Communities

RONALD M. ATLAS

1. Introduction

As used by microbiologists, the term *diversity* has various meanings, often describing qualitative morphological or physiological variances among microorganisms (Starr and Skerman, 1965; Belser, 1979; Hamada and Farrand, 1980; Hanson, 1980; Stanley and Schmidt, 1981; Walker, 1978; Yeh and Ornston, 1980). Microbial populations indeed exhibit great heterogeneity or diversity in their morphological, physiological, and ultimately genetic characteristics. An extensive list of diversifying factors that act to establish differentiating characteristics between microbial species has been discussed by Starr and Schmidt (1981). Some examples of these diversifying features are listed in Table I. These diversifying features have traditionally been employed by bacteriologists as the criteria for differentiating species. Often, the ability to recognize and distinguish species of microorganisms is difficult, but it is essential for assessing diversity.

In its ecological sense, the term *diversity* describes the assemblage of species within a community, and it is in this restricted sense that ecologists synonomously describe ecological and species diversity (Margalef, 1979; Pielou, 1975; Whittaker, 1975). Species diversity is a measure of entropy (disorder or randomness) of the community; an index of diversity measures the degree of uncertainty that an individual picked at random from a multispecies assemblage will belong to a particular species within that community (Legendre and Legendre, 1982). The greater the heterogeneity of the assemblage of populations and individuals within those populations, the greater the diversity of the

RONALD M. ATLAS • Department of Biology, University of Louisville, Louisville, Kentucky 40292.

Table I. Diversification Elements of Prokaryotes

Morphological

Cellular size
Cellular shape
Cellular flexibility vs. cellular rigidity
Morphogenesis and life cycles
Endospores, cysts, conidiospores, and
   sporangiospores
Cell division, binary fission, budding,
   fragmentation
Filaments, trichomes, multicellularity
Cytoplasmic inclusions and vacuoles
Mesosomes, thylakoids, and
   chromatophores

Cell envelope: gram staining and correlated
   properties, membrane diversity (lipopoly-
   saccharide, protein, lipid), wall (and pep-
   tidoglycan) diversity
Flagella and other locomotor devices
Prosthecae and other cellular appendages
Noncellular appendages, including pili and
   similar structures
Holdfasts and other adhesive devices
Sheaths, capsules, zoogleae, other extra-
   wall structures (including encrustation by
   iron and manganese oxides)

Associative relationships

Colonies, clones, multicellularity
Cooperation and competition

Symbiotic associations including endosym-
   biotic relationships
Host cell–parasite interactions

Physiological

Relationship to oxygen
Nutrition, mechanisms of energy conversion, metabolic potentialities—catabolic
   capabilities, growth factor requirements,
   dinitrogen fixation and other nitrogen
   metabolism
Pigments
Secondary metabolites
Luminescence

Relationship to light (photosynthesis, other
   phototransducers, other effects of visible
   light)
Relationship to radiation: UV, X-rays, and
   other ionizing radiation
Temperature relationships
Hydrogen ion concentrations
Barophily
Halophily
Osmotic relationships

Genetic and epigenic

Nucleoids, including genome size, multi-
   nuclearity, peculiar karyology
Plasmids, episomes, temperate phages,
   other nonnucleate elements containing
   nucleic acid

Ribosomal RNA sequencing
Isoenzymes

community. Diversity is equivalent to a measure of variance for the species parameter of the community; it is a measure of the species composition of an ecosystem in terms of the number and relative abundances of the species.

There are several assumptions inherent in assessing species diversity. The first is that populations occupying a particular habitat initiate interrelationships that result in the establishment of an organized community. There is ample evidence that biological populations, including microbial populations, establish interpopulation relationships that lead to the formation of a defined and stable community structure. Microbiologists have long recognized the "normal microflora" associated with various ecosystems, e.g., the normal microbiota of man. The occurrence of a normal microbiota must result from the establishment of community structure. Population (species) interactions that lead to the establishment of a defined community have been assumed to be based on various physiological interactions. The functional roles of specific populations (niches) within the communities of certain ecosystems have been defined, e.g., for the rumen ecosystem (Hungate, 1975); the bases for interspecific population relationships within such communities have been defined as well, leading to a relatively complete understanding of community structure and ecosystem function. Work using chemostats has elucidated some of the interactions between microbial populations that lead to the establishment of stable community structure in aquatic ecosystems; the nature of these interactions has been summarized by Slater (1978, 1980). In chemostat studies, it is often found that stability occurs when several interacting populations cooperate to best exploit the available resources. In some cases, two species constitute a stable community structure, whereas in other experiments, additional member populations are needed before community stability is achieved.

Whereas interrelationships among populations are clearly dynamic, the assumption that the populations within the community reach points of stability is an underlying principle of diversity calculations. In fact, perhaps the premise of greatest importance in considering ecological diversity and community structure is that species diversity is a community parameter that relates to the degree of stability of that community, i.e., that stable and resilient biological communities must contain a certain level of diversity; it is this hypothesis that is often used to justify the estimation of community diversity (Pielou, 1975; Peet, 1974). Communities with too much or too little diversity would be subject to continuous or catastrophic change.

2. Measurement of Species Diversity

Before considering how diversity measurements have been applied to microbial communities, we must examine the meaning of diversity and how it is measured. Ecologists have developed several indices of species diversity. Several extensive discussions of ecological diversity indices have been published (Dennis and Patil, 1977; Hurlbert, 1971; Legendre and Legendre, 1982; Margalef, 1968; Peet, 1974; Pielou, 1966a,b, 1969, 1975, 1977; Whittaker, 1975; Woodwell and Smith, 1969). As discussed in these reviews, the meaning, interpretation, and proper use of particular indices are often the subject of controversy among ecologists. Hurlbert (1971), for example, argues that species diversity, as defined by a variety of indices, has no biological meaning; communities having different species compositions are not intrinsically arrangeable in a linear order on a diversity scale. Pielou (1975, 1977) points out the problems with the methods of calculating species diversity that make interpretation difficult. Peet (1974) also considers the use and misuse of diversity indices. It is not the intent of this review to settle these long-standing debates, but rather to consider how the concepts embodied in the measures of ecological diversity can be extended to microbial communities. The expression of ecological diversity as a species-diversity index is an outgrowth of information theory. Essentially, any diversity index must measure the heterogeneity of information stored within the component populations of the community; the species-diversity indices that have been developed aim to describe the way in which information is apportioned within the community.

2.1. Species Richness

In its simplest form, the species-diversity index simply represents a count of the number of different species found occurring together. Communities with many different species have high diversity and those in which few species are found have low diversity. The number of species (n) can be used as a measure of the biological richness (species richness) of a community (Patrick, 1949): diversity $(D) = n$. Margalef (1951) proposed a standardization of the number of species (n) by the number of individuals (N): $D = (n - 1)/\ln N$. Similarly, Odum et al. (1960) proposed using $n/\log N$ and Menhinick (1964) suggested using $n/\sqrt{N}$ as measures of species richness.

To overcome the problem of estimating the numbers of species when samples are not the same size, Sanders (1968) developed the method of species rarefaction. This method consists of calculating the number of species that the samples would contain if they were the same size. Sanders's original formula was corrected by Hurlbert (1971) so that one obtains the expectancy of a number of species (n') in a standardized sample of N' specimens from a nonstandard sample containing n species, N specimens, and N_i specimens in species i according to the formula

$$E(S) = E(n') = \Sigma \left[1 - \frac{\left| \begin{matrix} N - N_i \\ N' \end{matrix} \right|}{\left| \begin{matrix} N \\ N' \end{matrix} \right|} \right] \quad \text{where} \quad \left| \begin{matrix} N \\ N' \end{matrix} \right| = N!/N'!(N - N')!$$

The proper use of rarefaction has been considered by Tipper (1979) and Simberloff (1972). To facilitate the calculation of $E(S)$, Simberloff (1978) has developed a computer program for computing the expected number of species.

2.2. Dominance and Evenness

Whereas $E(n')$ indicates the expected distribution of species within the community, the concept of species richness alone, as measured by a simple species-richness index, does not account for the evenness or equitability with which species (bits of information) are distributed within the community. The way in which information is distributed within the community is an important component of the heterogeneity of the assemblage of biological populations. Communities can be dominated by individual populations even if the species richness of the community is high.

To assess the degree of dominance, a species-diversity index was developed by Simpson (1949). The Simpson index is expressed as the function $\Sigma p_i^2 = \lambda$, where λ represents the probability that any two individuals picked independently at random from the community will belong to the same species. The function λ is a measure of the expected commonness of an event; the probability that two randomly chosen specimens belong to the same species is a measure of concentration as proposed by Simpson. The Simpson index is the opposite of diversity: the greater the homogeneity of a community, the greater the chance two randomly picked individuals will be of the same species; i.e., the lower the heterogeneity (diversity) of the community, the higher the value for λ. When a community is dominated by a single species, λ is high and approaches unity. Conversely, when there are numerous species that are relatively evenly represented, λ is low and is the probability of selecting two different species at random.

Another procedure for measuring diversity, based on a geometrical approach, was proposed by McIntosh (1967). The McIntosh index measures uniformity (U) within the community; it is expressed as $U = (\Sigma N_i^2)^{1/2}$, where N is the total number of individuals in the collection (N_i is the total number of specimens in a species). The larger the number of species, the smaller the value of U; the maximum value of U occurs when the sample contains but one species. The diversity (D) of the community is inversely related to U, and a diversity index based on McIntosh's U can be expressed as $D' = N - U$. The McIntosh diversity measures species diversity on the basis of the evenness of the apportionment of populations within the community.

2.3. General Measures of Species Diversity—The Shannon Index

In contrast to the McIntosh uniformity index and the Simpson dominance index, a general information index measures both the species richness and equitability components of community diversity. The Shannon diversity index, known with slight mathematical variations as the Shannon–Weaver and Shannon–Wiener indices, is probably the most widely used index for measuring species diversity (Shannon, 1948; Shannon and Weaver, 1949). The Shannon index is expressed as $H' = -\Sigma p_i \log p_i$, where p_i is the proportion of the community belonging to the ith species. The calculation of H' was simplified by Lloyd *et al.* (1968), who developed the following log base 10 formula: $H' = C/N(n \log_{10} N - \Sigma n_i \log_{10} n_i)$, where C is 3.3219, N is the total number of individuals, and n_i is the total number of individuals in the ith species.

The Shannon index has the following properties: (1) for a given number of species (S), H' has its greatest value when $p_i = 1/S$ for all values of i; i.e., H' is at its maximum when there is a completely even distribution of species within the community; (2) H' is at its minimum value (0) when a community is composed of only one species; (3) the diversity, H', increases with species richness such that H' is greater for a completely even community with $S + 1$ species than for a completely even community with only S species; and (4) the diversity of the community measured with the Shannon index can be partitioned into different fractions; i.e., the individual diversity indices for component groups, such as taxonomic families or genera, can be added to determine the total diversity of a biological community. As stated by Legendre and Legendre (1982), a probabilistic interpretation of the Shannon index lends itself to a measure of uncertainty regarding the identity of a randomly chosen specimen from the community; this uncertainty is smaller when the community in the sample is dominated by a few species, in which case H' is low; the value of H' also diminishes when the number of species gets lower, which also diminishes the uncertainty associated with the identification of a randomly chosen specimen.

There are certain limitations to the use of the Shannon index (Pielou, 1969, 1975, 1977). The Shannon index is appropriately used only when examining communities that are sufficiently large so that removing samples in a census does not cause any perceptible change within the community. In cases where the community is small and fully censused to determine diversity, it is necessary to use another measure of diversity; in such cases, Margalef (1958) has proposed the use of Brillouin's index (Brillouin, 1962). Brillouin's index, expressed as $H = (1/N) \log (N!/\Pi n_i!)$, where N is the number of individuals in the whole collection and n_i is the number in the ith species, has also been proposed for use when one performs sampling without knowing whether the sample is representative and the sample is therefore best treated as a collection. Peet (1974) suggests that contrary to the Shannon index, Brillouin's measure of uncertainty is not a good measure of diversity and its use as a community descriptor is rarely necessary. A problem with both the Shannon index and the Brillouin index is that both species richness and evenness play a role in determining the value of the index. Quite different communities, as a consequence, can have the same index. It is sometimes necessary to assess the species richness and evenness components individually in order to understand the factors controlling the structure of a community and the reasons a community has attained a certain level of species diversity. Further, when addressing the question of community stability, one must consider whether species richness or equitability of species distribution establishes the level of entropy of the community.

It is possible to calculate an equitability index (J) by comparing the measured or estimated value of H with the theoretically maximal value of H that would obtain if the distribution within the community was completely even: $J = H/H_{max}$. The calculation of J, however, requires the determination of H_{max}, which properly can be achieved only if the total number of species is known. Another measure that describes the way in which information is distributed within the community is the redundancy index (Margalef, 1958; Patten, 1963), which is calculated according to the basic formula: $R = (H_{max} - H)/(H_{max} - H_{min})$, where H is the calculated diversity index, H_{max} is the maximal diversity possible for the number of species within the community, and H_{min} is the minimal diversity possible for that number of species. One problem with such diversity indices that has been discussed by Sheldon (1969) is their dependence on the species count.

An additional problem with the Shannon index, and similar diversity indices, occurs when one tries to perform statistical analyses, which are necessary to determine whether significant differences exist between communities with respect to the structure of the populations within the community. The problem is that the mathematics used in calculating the Shannon index does not permit a true estimation of error; thus, the variability of the measurement is not truly determined. The problem arises from the attempt to determine the diversity of a large community (a requirement for use of the Shannon index) from a sample; the result of sampling is that H' is a biased statistic (Pielou, 1966b, 1977). H' underestimates the true diversity of the community (H) by an amount approximately equal to $S/2N$, where S is the total number of species in the community and N is the number of individuals sampled. No correction can be made for the bias of H' unless S is known, which rarely if ever is the case. Statistical methods have been developed for comparing values of H' (Hutcheson, 1970; Adams and McCune, 1979), but performing such comparisons is difficult and rarely done; therefore, interpretation of H' when comparing communities is usually ambiguous and subject to criticism.

Yet another problem, of a practical rather than a theoretical nature, is that the sampled populations are often not chosen at random and independently drawn from the community (Pielou, 1977). This is a consequence of the natural patchiness of distribution of ecological communities. As a result of biased sampling, the diversity of the community estimated from a sample is almost always less than the true diversity for the whole community. Using the technique of species rarefaction, it is possible to estimate error and thus to compare diversities of differing communities.

Considering the different indices available for describing the structure of microbial communities, Tinnberg (1979) compared five common diversity indices to see which would be most suitable for the study of long-time changes in a phytoplankton community. Margalef's index of diversity (D = species richness) showed the same course during the year as did the number of species. This was also valid for Brillouin's, Shannon's, Simpson's, and McIntosh's indices when based on number of individuals. Brillouin's and Shannon's indices based on cell number differed from the diversity values obtained by Simpson's and McIntosh's indices depending on their different sensitivity to the two components of diversity, species richness and species evenness. Tinnberg states that diversity indices that include both species richness and evenness components (e.g., all except Margalef's) are recommended, but their different sensitivities to species richness and evenness components should not be overlooked; he concludes that the description of the phytoplankton composition should always be given together with specification of diversity indices.

3. Taxonomic Diversity

Having considered some of the indices that have been developed for describing ecological diversity, we will now examine some of the studies of microbial communities in which the question of species diversity has been addressed. Such studies of microbial community structure have been limited in large part by the taxonomic difficulties involved in defining microbial species. Ideally, species of microorganisms would be recognizable by their appearance as are species of plants and animals. Classic approaches to identification of bacterial species, though, require extensive biochemical testing in addition to morphological observation; this time-consuming process makes it difficult to identify the large numbers of bacteria necessary for calculating diversity indices. However, many photosynthetic microorganisms have characteristic morphologies that permit their identification to the genus or species level on the basis of morphology alone; the ornate structures of the diatoms make them particularly easy to identify in this way. In fact, the classic approaches to the classification and identification of algae have been based on morphological descriptions of field specimens, and thus it is not surprising that algal communities have been examined in terms of species diversity.

For purposes of this review, the blue-greens will be treated as blue-green algae, rather than according to their newly recognized proper taxonomic position as cyanobacteria, because the blue-greens are still classified under the botanical code and identified by the morphological appearance of field specimens; virtually all studies of diversity of phytoplankton consider the blue-greens as algae. The ability to recognize species in field specimens is advantageous in determining species diversity; consequently, most studies on the diversity of microbial communities have been restricted to diatom and phytoplanktonic communities.

3.1. Algal Diversity

3.1.1. General Ecological Principles

One problem with studying algae on the basis of identification of species according to morphological appearance in a field specimen is that it is not generally established whether the algae are living or dead. Such studies run the risk of examining the diversity of dead allochthonous organisms, rather than the active members of the autochthonous community. In a study of living and dead diatoms in estuarine sediments, Wilson and Holmes (1981) discovered that the percentage of dead diatoms can be large, that ratios of dead to living diatoms vary between sediment habitats, and that measurements of live species and diversity are lower than measurements that include live and dead members of the assemblage. They concluded that the understanding of environmental factors that influence or control species distribution and abundance would be improved if the occurrence of living taxa only is considered. The problem of separating living from dead microorganisms is difficult but essential for assessing the functioning of communities. If the concept of diversity is to have biological meaning, it must be an accurate measure of dynamic interaction among populations within the community and the resilience of those populations to fluctuations of environmental parameters.

Within the community, it is assumed that living populations play functional roles filling the available niches. Organisms that do not successfully fill a niche are eliminated (May, 1976; MacArthur and Wilson, 1963, 1967; Preston, 1962a,b; Whittaker, 1975). With respect to phytoplankton, however, there is a paradox (Hutchinson, 1959, 1961) regarding the principle of competitive exclusion (Gause, 1934; Patten, 1961), which predicts that no two species can occupy the same ecological niche at the same time and place. The vast diversity of phytoplankton observed in many aquatic environments presents an apparent contradiction to this principle because all phytoplankton compete for the same basic resources, and since the euphotic zones of most natural waters are relatively homogeneous, such coexisting plankters may be simultaneously occupying the same niche. Several hypotheses have been proposed to explain this paradox, including contemporaneous disequilibrium (Richerson et al., 1970), i.e., that patches of diverse and of monospecific plankton assemblages exist at the same time (contemporaneously), but are spatially separated in the same body of water. A difficulty with this explanation is its dependence on the development of monospecific blooms between periods of significant turbulent mixing. A different explanation was given by Kemp and Mitsch (1979), who created a general model to explain the paradox of plankton. Simulation results from their mathematical model support the hypothesis that physical turbulence in an aquatic system can mollify pressures between plankton populations and permit the coexistence of species competing for the same resources. Using a highly simplified model as a point of departure, Kemp and Mitsch developed a new model, explicitly incorporating physiological growth mechanisms, to investigate the effects of both advective and turbulent components of water movements on the growth of three competing phytoplankton species. In the absence of water motion, no two species were able to coexist, whereas under the hypothetical conditions of advection without turbulence (laminar flow), just two species were able to occur contemporaneously. Coexistence of all three species was achieved only with the addition of a random turbulence component to the model's hydrodynamic function. This general coexistence was observed only when the major turbulence frequency approached the turnover rate of phytoplankton populations. There may be a limited region of periodicities and magnitudes for hydrodynamic energy in which all phytoplankton can exist, and most natural aquatic environments may fall within this region. A coupling of physical and biological processes in nature may be influenced by the relative frequency of those processes. This study thus establishes a basis for forming stable heterogeneous (diverse) phytoplankton communities; effectively, the discontinuities in the environment allow for overlapping niches: the greater the number of niches, the greater the potential species diversity.

3.1.1.1. Relationship between Diversity and Biomass. Another area of general ecological investigation concerns the relationship between biomass (productivity) and diversity of phytoplankton communities. As discussed by Margalef (1961, 1963, 1967, 1968, 1979), dominance within phytoplankton communities is directly related to productivity and inversely related to diversity and stability. Diversity in a community is principally a mechanism that generates community stability; dominance is principally a mechanism that generates community productivity. The maturity of a community is reflected in both its diversity and productivity. Mature ecosystems are complex—the presence of a large number of species (high diversity) allows for numerous interspecific relationships. An ecosystem that has a complex structure, rich in information as reflected by high species richness, needs a lower amount of energy for maintaining such structure. This lower need for energy input to maintain the diversity of a mature ecosystem is reflected in a lower rate of primary production per unit biomass, whereas a stable level of diversity is maintained within the community.

The work of Revelante and Gilmartin (1980) supports the theory that there is a strong inverse relationship between diversity and productivity. This inverse relationship is especially pronounced when environmental changes favor rapid phytoplankton growth. For example, when plant nutrients associated with river discharge create bloom conditions, phytoplankton communities with high biomass and low species diversity develop. Conversely, with lower nutrient fluxes and decreased phytoplankton biomass, species diversity increases. The relationship of community diversity to productivity makes diversity measurements useful for understanding and describing eutrophication processes.

The hypothesis that the total phytoplankton biomass can predict phytoplankton community structure independent of its taxonomic composition was tested in a study by LaZerte and Watson (1981). In a 2-year study on Lake Memphremagog, Quebec, Canada, which exhibits a marked axial trophic gradient, 133 samples were adjusted to uniform count sizes, and the range of diversity numbers based on proportional biomass was calculated for each period. Biomass was a good predictor of evenness, but not of species richness. Since this prediction is independent of changes in taxonomic composition, species richness is more directly related to season and changes in taxonomic composition than to the evenness of distribution of taxa within the community. This study points out that species richness and evenness are distinct components of community structure. One component of the diversity of community can reflect biomass while the other does not. The total number of species in the community can be the same (identical species richness) even when one species opportunistically outgrows the other members of the community and becomes dominant.

In a study of species composition, diversity, biomass, and chlorophyll of periphyton in creeks and rivers in Oklahoma, Wilhm et al. (1978) found more taxa and lower densities in the upstream stations in the creeks than in the downstream stations; this resulted in greater species diversity values at the upstream stations. Species diversity was generally lower in the river than in the creeks, reflecting the lower number of taxa (lower species richness) and the presence of several abundant species (lower evenness). Chlorophyll a (algal biomass) was generally greater at the downstream stations in the creeks than in the upstream stations. Species diversity did not reflect extremes in physical or chemical conditions or both. Rather, diversity was inversely related to algal biomass, following the principle that high biomass usually results from overgrowth of a few successful populations within the community. Thus, when an algal bloom occurs, there is a great increase in biomass but a severe decline in species diversity due to a pronounced shift in the evenness of species distribution. The decreased species diversity and increased biomass of algal blooms are recognized consequences of eutrophication.

Bartha and Hajdu (1979) examined the species composition of phytoplankton in Lake Velense, Hungary, in three areas of different eutrophic levels of the lake. They found that rare species do not greatly influence the diversity index (calculated from the data using Shannon's formula) for the phytoplankton community and that the plant nutrients reaching the lake cause increasing dependency in diversity of the phytoplankton there. During a 2-year period, the increase in the nutrient supply led to a substantial decrease in diversity. According to their hypothesis, the diversity index increases with increasing eutrophication level in a certain domain, whereas after reaching a turning point, it has a decreasing tendency.

Both high biomass and dominance by a successful opportunistic species (low evenness/low diversity) characterize the algal blooms associated with eutrophication. Bogaczewicz-Adamczak (1978), in an investigation of the structure and production of planktonic algae in a eutrophic lake in Poland, found that the biomass development of one species resulted in the predominance of large colonial organisms that diminished the diversity of a community and restricted the production capacity of the remaining phytoplankton individuals. Sullivan (1981) collected diatoms seasonally from a monotypic stand of *Distichlis spicata* from a marsh in Mississippi in which the marsh surface was enriched with NH_4Cl and exposed to high light intensity by clipping the grass shoots. Nitrogen enrichment increased H' and S in the spring. Clipping greatly reduced species diversity (H') and the number of taxa in the sample (S) in all seasons except winter, but did not stimulate the growth of filamentous algae. Of the 111 taxa encountered, clipping eliminated 9 preexisting taxa and introduced 3 new taxa into the community. As with community diversity, effects due to clipping were more prevalent than those due to nitrogen enrichment. The specific combination of clipping and NH_4Cl enrichment had basically the same effects on community structure as did clipping alone. Nitrogen enrichment greatly stimulated the aerial yield of intact *D. spicata* stands and the regrowth of those clipped 3 months earlier.

3.1.1.2. Effects of Grazing and Predation on Community Diversity. The effect of grazing is another area of interest concerning community structure. Predator–prey relationships are one of the major types of interspecies interactions that establish and maintain community structure. If the grazer population recognizes and specifically selects particular prey species, grazing will lower the species diversity of the community; if the grazing activity is not discriminatory, the biomass of the prey will be lowered, but the diversity of the community may not be changed. Brawley and Adey (1981) studied the effect of micrograzers on an algal community structure in a coral reef microcosm. The effect of amphipod grazing on algal community structure was studied within a 75-liter refuge tank connected to a 6500-liter closed-system coral reef microcosm. When amphipods *(Ampithoe ramondi)* were absent or present in low numbers, a large biomass of mostly filamentous algal species resulted. The microalgae disappeared when amphipod density increased beyond approximately one individual per square centimeter of tank surface. The macroalga *Hypnea spinella* germinated in the system in association with amphipod tube sites. *Hypnea spinella* plants remained rare until filamentous species were eliminated by amphipod grazing. Feeding trials confirmed that *H. spinella* was protected from grazing by its size rather than by a chemical defense strategy. The *H. spinella* community observed was similar to the flora described on algal ridges where physical conditions exclude fish grazing. Amphipods and similar micrograzers are probably responsible for the algal community structure of these ridges.

Rees (1979) studied community development in two freshwater microcosms that were identical with regard to initial chemical composition and biological inocula, except that in one microcosm, mosquito fish *(Gambusia)* and herbivorous catfish *(Plecostomus)* were added. Three distinct communities developed in the tanks: a phytoplankton–zooplankton assemblage and two periphyton–zoobenthos communities associated with the sides and bottom of the tanks, respectively. Community development was monitored closely in the zooplankton–phytoplankton assemblage and to a lesser degree in the other two communities. Developmental and successional patterns were similar in both tanks. A major difference between the tanks was the timing of succession of the zooplankton and zoobenthos, due principally to *Gambusia* predation. One major drawback of the use of the tanks as an experimental analogue for lakes was the development of a periphyton growth that eventually overwhelmed the biomass of the system. The tanks displayed a degree of successional replicability, a large number of species, and a diversity of community development, supporting the idea that microcosms of this size may be useful as experimental systems for higher-level trophic manipulation and observation of life cycles not amenable to field studies.

3.1.2. Influence of Environmental Factors on Community Diversity

In addition to the interactions among populations within the community that determine the level of diversity, physical–chemical (abiotic environmental) factors determine the level of heterogeneity within the biological community occupying a particular habitat. Studies on species diversity have frequently been concerned with the effects of environmental parameters (e.g., temperature, light intensity, salinity, physical nature of the substratum, and various pollutants) on community structure. Samuels *et al.* (1979) emphasized the importance of multiple variable interaction, as opposed to individual limiting factors, in controlling the dynamics of phytoplankton population dynamics as reflected in species diversity.

In many of the studies on environmental influences on diversity, the composition of the community has been followed seasonally in an attempt to elucidate the natural changes in diversity that occur and the environmental factors that drive these changes. For example, Hulburt (1963) examined the seasonal variations in the diversity of phytoplanktonic populations in oceanic, coastal, and estuarine regions. He found that in the deep ocean north of Bermuda during spring and summer, conditions for growth are poor, but physical conditions, such as high salinity and great depth, are favorable to the marine phytoplankton. The dominant species in a sample constitutes a modest proportion of the cells counted, and a considerable diversity of species is observed. When very good growth conditions, resulting in extreme dominance, are coupled with an adverse physical environment, there is a reduction in diversity. These conditions are found in several New England estuaries, where the lower salinity would appear to be unfavorable to the marine phytoplankton and where extreme shoalness may strand the larger, more rapidly settling types of diatoms. Populations from the open ocean and coastal water off New York in the fall and winter typify an intermediate situation. Moderately good growth is associated with pronounced dominance, and diversity is low in very small samples of the populations. When a large number of cells is counted, however, there exists the considerable diversity that might be expected where the salinity and depth are favorable.

Lapointe *et al.* (1981) studied the spatial and temporal patterns of community structure, succession, and production of seaweeds associated with mussel-rafts in the Ría-de-Arosa, Spain, at four stations for 1 year. Community structure changed due to seasonal changes in abiotic factors (e.g., light, temperature), and an algal succession on new mussel-rafts from an initial ephemeral flora (e.g., *Ulva rigida, Enteromorpha* spp., and *Polysiphonia* spp.) during the fall and winter changed to a kelp-dominated flora in the summer. During the summer, *Laminaria saccharina* dominated the innermost estuarine station; *Saccorhiza polyschides* dominated the three outer stations. Because of the succession to a kelp-dominated community, species diversity and evenness decreased over time; species richness generally increased. The algal community that occurred during the fall, winter, and early spring had a relatively low biomass ($0.1-2.2$ g C/m^3).

Seasonal variations in the phytoplankton community structure of Delaware Bay were studied by Watling *et al.* (1979). The flora was composed primarily of small flagellates during the summer and early fall; diatoms dominated from October to May. Peak cell numbers occurred during the fall and early spring blooms. Evenness diversity was lowest during periods of maximum diatom abundance and highest when microflagellates dominated. There was a gradual shift in dominance, except during the early spring *Skeletonena costatum* bloom. Cluster analysis allowed the separation of the flora into three time groups and eight recurrent species groups.

In a study of diatoms in a thermal stream, Stockner (1968) found that despite seasonal changes in abundance and shifts in the species composition of the community, the values of the Shannon diversity index did not deviate significantly from the mean annual diversity value. The relative constancy of H' was interpreted as an indication that the diatom community was quite stable in this thermally heated stream. The variations in species composition within the community appear to be related to seasonal variations in light intensity. The annual succession of phytoplankton in a heated pond in central Finland was studied by Eloranta (1976, 1980). Phytoplankton biomass reached an initial maximum in March due to the growth of cryptophytes and *Asterionella gracillima*. The autumn maximum in September was due to growth of *Cyclotella meneghiniana*. Except in June, green algae dominated from April to

August. The average of the phytoplankton biomass for the whole year was 3.34 g C/m^3 and 4.44 g C/m^3 for the growing season March to September. The biomass minimum during the darkest winter in late December and January was less than 0.2 g C/m^3; the chlorophyll *a* concentration of phytoplankton fresh weight biomass varied between 0.2 and 0.7% (average about 0.34%). The maxima of the Shannon diversity (calculated with natural logarithms and on the biomass basis of each species) were at the time of maximum species richness in summer ($H' = 2.5$–2.9) and at the time of phytoplankton minimum in winter ($H' = 2.7$–2.8). No significant correlation occurred between Shannon diversity and the logarithm of the phytoplankton biomass.

An investigation of the structure and dynamics of the phytoplankton assemblages in Lake Kinneret, Israel, was carried out by Pollingher (1981), who reported that the phytoplankton of this warm monomictic lake is dominated by a Pyrrhophyta–Chlorophyta assemblage. Four stages of succession of phytoplanktonic algae occur in the lake, starting with thermal and chemical destratification and ending with stratification. The index of diversity of the phytoplankton communities is high during the destratification and mixed periods. The index reaches minimal values during late summer, when the ecosystem is subject to strong physical, chemical, and biological stresses. The diversity in Lake Kinneret increases with the increase in nutrients and not with the increase in temperature. Through most of the year, the nanoplanktonic forms are in greater numbers than the net plankton species. This fact correlates with the amounts of available nutrients in the light period. The annual average of the wet water trophic biomass in Lake Kinneret is very high in comparison with that in other warm lakes. The contribution of the nanoplanktonic species to the total algal biomass is very small during the *Peridinium* bloom, but represents about half the total algal biomass during the rest of the year. Concentrations of nutrients in the water, together with the adverse competitive effect of *Peridinium* or other algae, are largely responsible for the composition, succession, and abundance of the phytoplankton assemblages in Lake Kinneret.

Lakkis and Novel-Lakkis (1981) described the composition, annual cycle, and species diversity of the phytoplankton in Lebanese coastal waters. A total of 263 taxa were identified, of which 107 (45 genera) were diatom species and 157 dinoflagellates (25 genera); many of them were of Indo-Pacific origin as a result of the migration pattern through the Suez Canal. Annual and seasonal fluctuations were similar from year to year. Together with the associated planktonic diversity and abundance, maximum and minimum values of phytoplankton density were observed in May and December, respectively. The diversity index was highest in February and lowest in July. During summer, because of water stratification, high surface water temperature, strong light intensity, and nutrient depletion, low productivity was observed.

Sze (1980) studied the seasonal succession of phytoplankton in Onondaga Lake, New York. The seasonal succession of phytoplankton in the lake over a 2-year period was similar to the year period immediately after a reduction in the amount of phosphorus entering the lake. Diatoms and flagellates were dominant in the spring and replaced in importance by chlorococcalean green algae and the diatom *Cyclotella glomerata* during the summer. A die-off of summer algae was observed in the last year of the study. During the summer, silica was depleted in the epilimnion by diatom growths and phosphorus by chlorococcalean green algae. Enrichment studies indicated that the availability of phosphorus to green algae and silica to diatoms may have contributed to determining their periods of dominance. Blue-green algae continued to be relatively unimportant in the plankton, in contrast to the prestudy conditions, when they caused late summer blooms.

Sullivan (1978) performed taxonomic and statistical analyses of the diatom communities of a Mississippi salt marsh. Edaphic diatoms were collected on a seasonal basis from beneath five monospecific stands of spermatophytes in a bay marsh from fall 1976 to spring 1977. Of the 119 diatom taxa encountered, only 7 were restricted to a single edaphic habitat, and 5 of these accounted for 17.2% of the individuals comprising the community associated with *Distichlis spicata.* The single most abundant diatom was *Navicula tripunctata,* which accounted for 21.5% of all individuals. Based on a two-way analysis of variance of species diversity (H') and the number of taxa in a sample, edaphic diatom community diversity was highest beneath *D. spicata* and *Spartina patens,* lowest beneath *S. alterniflora* and *Juncus roemerianus,* and somewhat intermediate for the *Scirpus olneyi* habitat. Structural differences between selected community pairs were quantified using a similarity index, and the values generated were exceedingly variable. A multiple regression analysis revealed that structural differences among edaphic diatom communities were related to differences in elevation, far-red light energy, ammonia N, soil moisture, and, tentatively, height of the plant canopy.

Fagerberg and Arnott (1979) studied seasonal changes and structure of a submerged blue-green algal–bacterial community from a geothermal hot spring. The species structure and seasonal dynamics of a thermophilic blue-green algal–bacterial community from Mimbres, New Mexico, were described; with the use of electron-microscopic techniques that preserve three-dimensional relationships (stereological methods), several common quantitative ecological parameters (biomass, density, distribution measurements, species-diversity indices, and importance values) were described for the community growing about 1 m beneath the water surface of the inner walls of a concrete cistern. Temperature and pH were relatively constant, whereas water flow rates, duration, and intensity of light showed seasonal variation. Species-diversity indices, density, and importance values for the blue-green algal and bacterial populations showed seasonal changes. The three dominant blue-green algal species that represented the greatest amount of biomass alternated in their abundance on a seasonal basis, whereas the three dominant bacterial cytospecies remained the same throughout the period of study. Considerable seasonal change occurred in the species structure of this community, but the total living volume remained relatively constant, with the major changes in the blue-green algal populations. Many seasonal changes observed appeared to correlate with seasonal changes in light intensity and duration.

Studies of changes in the community structure of a geothermal algal–bacterial community were carried out by Fagerberg and Arnott (1981) utilizing electron microscopy as an analytical device. Quantitative information was obtained from micrographs of algal mats, permitting a description of the mat community in terms of its biomass, organism-distribution indices, organism density, and operational taxonomic-unit diversity. An important feature of analyzing microcommunities in this way is that the spatial integrity of the organisms within the community is retained, while considerable quantitative information is acquired. The submerged community, consisting of cylindrical strands (about 90 mm long) growing in a fast-flowing, hot-water stream (54°C) in Mimbres, New Mexico, contained primarily blue-green algae (cyanobacteria), bacteria, and amebae embedded in a gelatinous matrix. Blue-green algae were distributed throughout the entire community, but were more concentrated at the periphery. The central portions of the strand contained blue-green algae, but were dominated by bacteria; thus, streamer communities were abundant in the stream throughout the year, but biomass, organism density, and distribution of organisms within the mat changed during the four sampling periods, while diversity indices remained constant.

Stevenson and Stoermer (1981) found that algal community structure in Lake Michigan was different in shallow, mid-depth, and deep zones. In shallow zones, the diatom communities had high diversities and low abundances, which was probably due to substantial substratum disturbance by wave action in this zone. High abundance and high diversity in mid-depth communities probably occurred because lower wave disturbance allowed algal accumulation in this zone. Low abundance and low diversity characterized deep-water communities, where planktonic algae accumulated and low light levels reduced growth of benthic species.

Moore (1981) studied benthic algae in littoral and profundal areas of a deep subarctic lake. The number of species associated with sediments declined with depth. *Achnanthes minutissima, Navicula pupula, Cymbella* spp., and *Nostoc pruniforne* reached greatest relative abundance in shallow water, whereas *Nitzschia palea* and *N. dissipata* were relatively common below 20 m. *Amphora ovalis, Gyrosigma spenceri,* and *Tabellaria flocculosa* did not exhibit a consistent distribution pattern. Low light levels and, to a lesser degree, temperature were the most important factors influencing the diversity, species composition, and density of the epipelon in deep water. Although *Achnanthes pinnata* was rare in deep water, the relative abundance of other common species (*Amphora ovalis* var. *pediculus, Fragilaria construens* var. *venter*, and *Achnanthes minutissima*) was constant at all stations. Although low light levels probably controlled the densities in deep water, the physical characteristics of the substratum determined the diversity and species composition of the community. Moore (1979) also studied the factors influencing the species composition and diversity of phytoplankton in 21 arctic and subarctic lakes. The maximum number of species, 60–75, and highest diversity indices, 1.6–1.9, were recorded from the most southerly lakes situated at 63–64°N. Since there were fewer species (40–60) and lower diversity indices (0.7–1.0) in lakes located at 65–66°N, temperature was concluded to be the main factor influencing the diversity of the flora. Other parameters, including surface area, depth, and the concentrations of nutrients, exerted little influence on the plankton. The maximum densities of several common species increased significantly in lakes with high concentrations of total phosphorus and total hardness, whereas the reverse pattern was recorded for two other common species. Diatoms were restricted primarily to large lakes in which turbulence helped the algae float.

Anderson *et al.* (1981), in a study of diatoms and dinoflagellates in Hudson Bay, Canada, found various patterns of distribution for diatoms and dinoflagellates. Some algal species were found only inshore, whereas others were distributed throughout the bay, with the exception of the low-salinity waters of the southwest coast. The lowest diversities of diatoms, dinoflagellates, and diatoms and dinoflagellates combined were observed at lower salinities along the southwest coast. The diversity index of diatoms and dinoflagellates combined was otherwise high throughout the bay ($H' > 3$) and was highest in the Coats and Mansel Islands area. Bruno *et al.* (1980) found a mean Shannon diversity index of 1.27 for the phytoplankton community of Peconic Bay. The generally higher diversities at the outer portions of the bay were attributed to the incursion of seawater. At a few locations, e.g., at the headwaters during the spring, the diversity was low due to large algal blooms.

3.1.3. Colonization and Succession

Various investigations have been concerned with the colonization of substrata by diatoms. The basic premise of these studies is that the nature of the processes of substratum invasion and the successional process that follows determine the composition of the community. Patrick (1963) performed some of the earliest studies that employed diversity measurements to examine algal community structure. In one of these early studies, Patrick (1967) carried out a series of experiments to investigate (1) the effect of area size and the number of species in the species pool, which were capable of invading an area, on the number of species that composed the diatom community; and (2) the effect of invasion rate on the structure of the community. Additionally, this investigation examined the structure of diatom communities in differing streams of similar temperature and chemical composition. She concluded that the size of the area, the number of species in the species pool capable of invasion of the substratum, and the invasion rate all influence the species richness and overall diversity of the diatom community. A reduced invasion rate (size of area and number of species in the species pool remaining the same) lowered the total number of species in the community, particularly those species that were normally present in low numbers within diatom communities. This decrease in species diversity was seen as a lowered species richness (Fig. 1). Species with fairly large populations increased, but the total number of species decreased.

One of the main results of high invasion rate is to maintain within the community species with relatively low numbers, i.e., to ensure the presence of rare species. Such rare species increase diversity and may act to stabilize a community under variable environmental conditions. If environmental conditions change, such rare species may be better adapted to survival than the presently common species, and the species composition, especially relative abundances, of the community might shift. The size of the area available for invasion affects the composition of the community. As species invade the area, the rate of increase of species richness declines, but, as found by Patrick, increasing numbers of species within the community need not alter the biomass of that community. Eventually, the number of species reaches a level of stability. Theoretically, once an area becomes filled, the first species to be eliminated are those represented by small populations. Although exclusion of diatom species on substrata resulting from interspecific competition has not been proven, most evidence supports the idea that competition determines which organisms are retained within the community.

In later studies, Patrick (1976, 1977) found that at first the diatom community is two-dimensional. When growth becomes relatively heavy, many of the species that formerly lay flat on the substratum stand up, giving rise to a three-dimensional community structure. In a more recent study, Hudon and Bourget (1981) considered the initial phases of community development of an artificial substratum using scanning electron microscopy. This study concluded that although the occurrence of stalked forms is apparently dependent on the

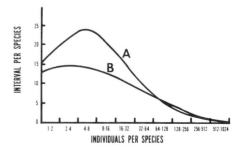

Figure 1. Effect of invasion rate on diatom diversity. (A) Invasion rate = 550–600 liters/hr; at height of mode, diversity index = 3.8 and number of species = 22; (B) invasion rate = 1.5 liters/hr; at height of mode, diversity index = 1.0 and number of species = 15. After Patrick (1963).

presence of flat-lying species, the sequence of dominance always proceeds toward vertical species. This study further recognized the importance of bacteria in establishing a primary film that is critical for the adsorption of diatoms and the establishment of the structure of the diatom community. The ability of the artificial substrata to retain water through detritus and surface irregularities is probably the main factor controlling the development of the diatom community. Stevenson (unpublished manuscript) examined seasonal variability and substratum specificity of diatom species diversity in the Sandusky River, Ohio. He found that there was a downstream increase in diatom diversity, in some habitats, during the winter and spring, but not during the summer and fall. On the basis of field observations, Stevenson suggested that fluctuating water levels in winter and spring disturbed diatom communities and established a diversity gradient; changes in diversity were correlated to colonization time since the last disturbance.

Stevenson (1981, 1983) proposes that recognition and delineation of successional mechanisms, including immigration, reproduction, emigration, death, and grazing, are necessary to determine the manner in which environmental factors control the diversity and species composition of diatom communities. Benthic diatom communities in areas sheltered from the main force of current in streams have higher abundance, more species, and lower evenness than communities exposed to direct current. According to Stevenson, species richness of diatoms in sheltered habitats should be greater than in exposed habitats because immigration rates are negatively related to exposure and current stress is positively related to exposure of substrata. As new species migrate into a community, species richness increases; this process is more rapid with faster immigration in protected areas. Also, Stevenson proposes that evenness of communities exposed to current can be lower than in sheltered areas because current enhances reproductive potentials of benthic algae. Evenness decreases during development of the community when a few species reproduce more rapidly than others. According to Stevenson's experiments, emigration, grazing, and death of diatoms are not important factors in establishing community diversity; rather, diversity in the community depends on the immigration rate and the reproductive capacity of the diatom populations (Fig. 2).

Eloranta and Kunnas (1979), in a study of the growth and species communities of the attached algae in a river system in central Finland in which water quality and the accumulation and growth of algae on the artificial substratum were followed, found that the growth of diatoms depended on the speed of water flow, turbidity, and phosphorus content of water. Two algal maxima, in the spring and autumn, were noted in clear-water streams, but in some streams with turbid water, one or both of these maxima were absent, notwithstanding the relatively high nutrient concentration of the water. An important conclusion of this study is that following initial recruitment of species (immigration), species diversity of diatom communities decreases as they

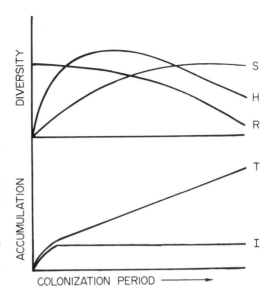

Figure 2. Changes in diversity of diatom community during colonization period. (*S*) Species richness; (*H'*) Shannon diversity index; (*R*) redundancy index (measure of evenness); (*T*) rate of total accumulation of diatom; (*I*) immigration rate. The area of the curve between *T* and *I* represents accumulation due to reproduction. After Stevenson (1983).

develop. The age of a diatom community is thus reflected in the species diversity of the algae in that community.

The studies on substratum colonization indicate the importance of the initial populations and the processes controlling the successional changes in species composition that lead to formation of the stable community. Sousa (1979a) investigated mechanisms of ecological succession in field experiments in a rocky intertidal algal community in southern California. The study site was an algal-dominated boulder field in the low intertidal zone. The major form of natural surface disturbance, which clears space in the system, is an overturning of boulders by wave action. Algal populations recolonized cleared surfaces either through vegetative regrowth of surviving individuals or by recruitment from spores. Boulders that are experimentally cleared are colonized within the first month by a mat of the green alga *Ulva*. In the fall and winter of the first year after clearing, several species of perennial red algae colonize the surface. If there is no intervening disturbance, *Gelidium canaliculata* gradually dominates the community, holding 60–90% of the cover after 2–3 years. During succession, diversity increases initially as species colonize a bare surface, but declines later as one species monopolizes the space. Selective grazing on *Ulva* by the crab *Pachygrapsus crassipes* breaks this inhibition and accelerates succession to a community of long-lived red algae. Grazing by small mollusks, especially limpets, has no long-term effect on the successional sequence; their grazing temporarily enhances the recruitment of the barnacle *Chthamalus fissus* by clearing space in the mat of algal sporelings and diatoms that develops on the recently denuded rock surfaces.

Small boulders with a shorter disturbance interval support only sparse early-successional communities of the green alga *Ulva* and barnacles (Sousa, 1979b). Large, infrequently disturbed boulders are dominated by the late-successional red alga *Gigartina canaliculata*. Intermediate-size boulders support the most diverse communities composed of *Ulva,* barnacles, several middle-successional species of red algae, and *G. canaliculata.* Comparison of the pattern of succession on experimentally stabilized boulders with that of colonies found on unstable ones confirms that it is the differences in frequency of disturbance that are responsible for the aforementioned patterns of species composition. The frequency of disturbance also determines the degree of variation among boulders in species composition and diversity. Small boulders that are frequently overturned sample the available pool of spores and larvae more often, and so a greater number of different species occur as single dominants on these boulders. Boulders with an intermediate probability of being disturbed are most variable in species diversity. Assemblages on these boulders range from being dominated by a single species to being very diverse, whereas most communities on boulders that are frequently or seldom disturbed are strongly dominated. Observations on the local densities of three species of middle-successional red algae over a 2-year period indicated that most of these are variable in time. More local populations were extinct or became newly established on boulders and remained constant in size. These species persist globally in the boulder field mosaic by colonizing recent openings created by disturbances. The results suggest a nonequilibrium view of community structure and disturbances that open space for the maintenance of diversity in most communities of sessile organisms.

Grimes *et al.* (1980) compared epiphytic diatom assemblages on living and dead stems of the common grass *Phragmites australis*. Diatoms epiphytic on *P. australis* were collected from a single clone at the southern end of Provo Bay, Utah Lake, Utah. Diatom populations of both living and dead stem sections were analyzed. The species diversity in each sample was high, indicating that the stems provided a relatively stable habitat for diatom epiphytes.

Drifting algae represent a different type of substratum that is subject to colonization by algae, i.e., a different habitat in which algal succession occurs. Yoshitake (1981) found, in a study of drifting algae in a river at eight stations in the middle and lower regions of the Tama River in central Japan, an abrupt increase of algal density in the main course of the river. The first and second dominant species were composed of planktonic species or attached algae that were also capable of living as plankton. In this case, the diversity indices based on Shannon's formula were generally low because the drifting algal communities were confined to several species. Such an abrupt increase of the algal density with a low diversity value suggests that the dominant species might derive from proliferating planktonic algae rather than removal of attached algae. Compared with the average cell density in Japanese rivers, higher values were found at a few stations of the tributary streams. The first and second dominant species of these stations consisted of attached algae that have a tolerance for water pollution, and in addition, the diversity indices were low (1.10–2.62). Negative correlation was recognized between diversity and total cell number and between the dominance index and the diversity index. High values for dominance were recognized at stations where drifting algae increased profusely or the waters were polluted.

3.1.4. Effects of Environmental Stress on Community Diversity

In cases of severe environmental stress, such as the input of high concentrations of toxic pollutants, species diversity generally declines. Patrick (1963) found that algal diversity was lower in polluted streams than in relatively pristine waterways (Fig. 3). Marshall *et al.* (1981) carried out an *in situ* study of cadmium and mercury stress in the plankton community of a lake in Northwestern Ontario, Canada. Significant changes in lake plankton communities were observed as a decrease in community diversity at molar concentrations of less than 0.2×10^{-8} M/liter for both Cd and Hg. Potentially toxic concentrations of Cd may occur in many contaminated lakes, but toxic concentrations of Hg probably occur in few lakes.

Eloranta and Kettunen (1979) examined the impact of discharges from a paper mill. Near the cellulose factory, the Shannon diversity index and the evenness within the community were found to be high, but in eutrophicated areas, only a few species were dominant and both evenness and diversity were low. Nelson *et al.* (1976) found that addition of an insecticide to soil lowered diatom diversity, indicating some diminution of environmental quality. The decline in diversity was not detected until 12 weeks following addition of the insecticide. Stockner and Benson (1967), in an examination of succession of diatom assemblages in Lake Washington, found a high correlation between community structure and sewage input. In sediments not impacted by sewage, the diversity of the diatom community remained fairly constant. The input of sewage was reflected in a change in dominance within the community; e.g., *Melosira italica* was replaced by *Fragilaria crotonensis* in some impacted sed-

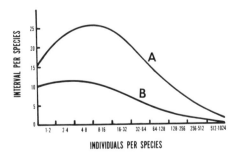

Figure 3. Effect of pollution on the diversity of diatom communities. (A) Diatom population from Ridley Creek, Pennsylvania, a nonpolluted waterway; (B) diatom population from Lititz Creek, Pennsylvania, a polluted stream. After Patrick (1976).

iments. The cultural enrichment due to sewage input thereby caused a considerable change in species importance rankings and species diversity indices.

Although the diversity of a community often decreases in response to the input of acute levels of pollutants, this is not always the case. The response of a community to the input of pollutants, as reflected in diversity measurements, depends on the factors controlling the natural distribution of the species within that community. Consequently, while the input of pollutants, or any severe environmental stress, invariably alters the environmental stress on a community, diversity sometimes decreases and sometimes increases. For example, Saifullah and Moazzam (1978), in a study of marine centric diatoms from the highly polluted lower harbor in Karachi, Pakistan, found that the temporal and spatial distributions of species were irregular and species diversity was high. They found that the species composition was significantly different from that of the shelf and speculated that the high species diversity may be the result of the extremely unstable environmental conditions in the harbor.

Similarly, Thompson and Ho (1981), studying the effects of sewage discharge on phytoplankton in Hong Kong, found no evidence of a reduction in taxonomic diversity in polluted areas except in the summer, when the net phytoplankton was dominated by *Chaetoceros* spp. In the autumn and early winter, *Skeletonema costatum* was abundant in the central polluted areas. The coastal waters of Hong Kong constitute a transition from estuarine conditions in the west to more oceanic conditions in the east, with the major discharge of untreated sewage located at the midpoint. Stevenson and Stoermer (1982) likewise found high diversity in an area of Lake Huron that was receiving inputs of sewage effluent. They suggested that species diversity is a curvilinear function of growth inhibition. According to their hypothesis, low species diversity is observed in slow and rapidly growing algal assemblages and high species diversity may occur in assemblages growing at a moderate rate. Species diversity could thus be greatest at some moderate level of stress perturbation, especially if growth of a dominant species is inhibited, leading to enhanced evenness of species distribution within the community. At high levels of growth inhibition, species richness is probably lower than at moderate levels of growth inhibition. Therefore, a group of organisms in a highly stressed environment would have a lower diversity than a group in a moderately stressed environment because of the elimination of sensitive species.

Patrick *et al.* (1954) concluded, on the basis of numerous river surveys, that in a river that has not been adversely affected by pollution, conditions are favorable for the occurrence of many species within the community. In unpolluted habitats, the majority of species are represented by a relatively small number of specimens; competition is strong and many habitats are present, each occupied by a different species. The effect of pollution is to eliminate the more sensitive species; competition is reduced and the more tolerant species proliferate. Thus, there are fewer species with more individuals in each, unless the stress is too severe, in which case the entire algal community may be eliminated. These studies emphasize the importance in certain situations of considering evenness and species richness as separate parameters in order to understand the reason for a particular state of diversity of a community.

3.2. Protozoan Diversity

Like algae, many protozoan species are recognized on the basis of morphology. Even so, protozoan communities have not received the same level of attention as algal communities. Hatano and Watanabe (1981) studied seasonal changes of protozoa and micrometazoa in a small pond with leaf litter supply over a 2-year period. The total number of taxa observed in this pond was 83 protozoans and 30 micrometazoans. Individual taxa exhibited habitat preferences; some species occurred primarily in surface water, others on leaf litter on the pond bottom, and still others on sedimented mud. The microfauna in the litter on the pond bottom had a higher diversity than in the water or the sedimented mud. Patterns of seasonal change in the density of organisms, with one or two peaks in a year, were recognized in some taxa of protozoa. Most peaks of bacteriovorous protozoa in the leaf litter appeared in the late autumn and spring, a phenomenon considered to be closely related to the litter decomposition because of the reduction of dissolved O_2 or high bacterial density, or both, that occurs during these two seasons. Foissner (1980) studied species richness and structure of the ciliate community in small water bodies in the Austrian Alps, 1150–2600 m above sea level. Species and individual richness and group dominance were similar to those of natural small water bodies in other areas of the world. Bacterial consumers and omnivorous species were dominant; thus, the species–individual relationship and the genus–species relationship obeyed the rules known from metazoan communities; the data support the contention that these alpine pools are not extreme environments for ciliates.

3.2.1. Effects of Predation on Community Diversity

Several investigators have used protozoa as experimental organisms to examine the influence of predator–prey interactions on community structure. In one such study, Hairston et al. (1968) examined the relationship between species diversity and stability using an experimental approach with protozoa and bacteria. They used small experimental communities of bacteria and protozoa to test the widely held hypothesis that higher species diversity brings about greater stability. Three species of bacteria, three species of *Paramecium,* and two species of protozoan predators, *Didinium* and *Woodruffia,* were used. The communities were maintained by regular additions of appropriate combinations of species of bacteria. Stability was measured as persistence of all species and as a tendency to maintain evenness of the species-abundance distribution; these two measures of stability were in general agreement. Stability at the *Paramecium* trophic level was increased by increasing diversity at the bacterium level, but three species of *Paramecium* were less stable than two species communities. An important finding was that one pair of *Paramecium* species consistently showed greater stability without the third species than with it, indicating that there were significant second-order effects, with two species having an interaction that was detrimental to the third species. They concluded that much more experimental work is needed before the functional relationships between diversity and stability are defined.

In a more recent study, Luckinbill (1979) examined the stability, regulating factors, and dynamics of an experimental community of protozoans. The predator *Didinium nasutum* coexists in microcosms with its prey, *Colpidium campylum*. Prey, including *Paramecium depraurelia* and *P. primaurelia*, was limited by the quantity of bacteria and nutrients available. *Didinium,* an ineffective predator, was limited by the availability or quality of prey as food, or by both. Enrichment of this community with bacteria or nutrients resulted in the extinction of prey and starvation of the predator. Increasing diversity of the lower trophic level by adding alternative prey species destabilized this community, also causing extinction. Stability depended here on the characteristics of the particular species serving as prey, not simply on the diversity of species present.

Kuserk (1980) studied the relationship between cellular slime molds and bacteria in four soils. [The cellular slime molds are considered as protozoa in the latest classification of protozoa (Levine *et al.,* 1980) and are therefore discussed here.] Observations of natural and manipulated populations were used to investigate the regulation of population sizes and community structure in a group of soil ameba, the cellular slime molds *(Dictyostelium)*. Correlations of slime mold abundance and distribution patterns with soil bacteria in the field suggest that food supply is a potential factor in the regulation of these species numbers. One species, *Dictyostelium mucoroides,* responds most visibly to seasonal changes, spring and fall being the peak seasons. When total slime mold numbers are partitioned into active and encysted forms, the ameba account for as much as 51 and 24% of the population in the fall and spring, respectively, vs. only 10–12% of the population during the summer and winter months. Large additions of various bacteria to field plots cause significant increases in *D. mucoroides* numbers. More often, the ability of the species to respond to a second addition of bacteria, made several days later, depended on its density. High-density populations failed to respond to additional food, whereas those that had already returned to base levels showed increases. These findings support the hypotheses that cellular slime molds are food-limited in nature and that community diversity is due, at least in part, to differential resource utilization by these species in nature.

3.2.2. Colonization of Substrata and Succession

Experiments with protozoa have been performed to investigate the factors influencing protozoan colonization of substrata and the effects of initial colonization on the eventual structure of the community. Henebry and Cairns (1980a,b) exposed artificial substratum islands to source-pool protozoan communities of differing maturities (i.e., some stage between pioneer and mature) in a series of microcosm experiments. Exposure of initially barren islands to source pools that were in an early state of development resulted in significantly more rapid initial colonization rates and faster attainment of equilibrium species numbers than exposure to the most mature source pools. Their results supported the hypothesis that rates of colonization onto the islands were influenced by the maturity of the source pools (state of diversity as determined by successional stage) and the proportion of pioneer species in the source-pool communities. A hypothesis supported by this study is that species composition of source pools may be of greater importance than the size of a nearby source pool in determining the rate of recovery of a community from pollution stress.

3.2.3. Effects of Environmental Stress on Community Diversity

Cairns (1969), in an examination of factors affecting the number of species in freshwater protozoan communities, observed a decrease in diversity following a temporary thermal shock; within a few days, diversities of the thermally stressed communities returned to those of non-thermal-stress controls (Fig. 4). Cairns et al. (1979) have proposed the use of protozoan communities developing on artificial substrata for the assessment of pollutional stress. They point out that stress distorts the distribution of species within the community by eliminating many scarce to moderately abundant (sensitive) species and by increasing the numbers of abundant (tolerant) species. The net result is that a

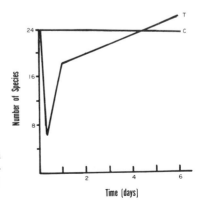

Figure 4. Effect of thermal shock on a protozoan community. After a rapid initial decline, the diversity of the community subjected to shock (T) returns to control (C) levels. After Cairns (1969).

stressed community has a greater proportion of its total species present in high abundance, i.e., a lower diversity than it would have if it was not under environmental stress. Such changes in community structure must affect the dynamics of the initial colonization of barren islands, drawing initial colonizers from the pool of species in a stressed community. The attainment of a stable equilibrium number should be more rapid for artificial substratum islands receiving colonists from a pollution stressed species pool, in part because there is generally a higher incidence of opportunistic colonizers in such stressed communities and in part because the eutrophic conditions often associated with pollution stress make the habitat favorable for high growth rates. Accordingly, the colonization process of barren islands by protozoa and the dynamics of establishing a stable community at some level of species diversity are amenable to experimental determination and may be a useful index of environmental stress due to pollution.

3.3. Bacterial Diversity

The problems in measuring species diversity of bacterial communities inherent in the identification of bacterial species have been discussed by Staley (1980). Because of the difficulty of identifying bacterial species directly on the basis of their morphological or colonial attributes, the measurement of heterotrophic bacterial diversity is still novel when compared to species-diversity measurements for plants, animals, and algae. Despite this difficulty, several recent studies have used diversity indices to describe bacterial communities. Swift (1976) has considered the relationship of the state of diversity of the microbial community to decompositional processes. Several investigators have examined the diversity of bacterial populations involved in the biogeochemical cycling of nitrogen. Belser and Schmidt (1978) studied diversity in the ammonia-oxidizing/nitrifying populations of a soil. Multiple genera of ammonia-oxidizing chemoautotrophic nitrifiers in a soil were detected, isolated, and studied by modified most-probable-number (MPN) techniques. Belser and Schmidt examined a Waukegon silt loam soil that was treated with ammonium nitrate or sewage effluent. The genera *Nitrosomonas* and *Nitrosospira* occurred more commonly than the genus *Nitrosolobus*. Three different MPN media gave approximately the same NH_3-oxidizer counts within statistical error after prolonged incubation, but differed markedly in ratios of *Nitrosomonas* to *Nitrosospira*. Sepers (1981) examined the diversity of the ammonifying microorganisms in natural waters. A single linkage clustering demonstrated that the bacterial population, which is able to take up amino acids in natural waters, is composed of a variety of microorganisms that differ only slightly in their ability to utilize a variety of organic compounds. Gamble *et al.* (1977) used the Simpson index of dominance and the McIntosh index of evenness in a study of denitrifying populations in soils to describe the numerically dominant denitrifying bacteria. They found that there was no correlation between the diversity of denitrifiers and soil properties. Also, the indices of diversity did not correlate significantly with numbers of denitrifying bacteria in soil.

Woodruff (personal communication) has examined the diversity of actinomycetes in Australian soils. Evidence was obtained in his studies to indicate that actinomycete populations of soils and of rhizosphere areas are relatively stable with time, unless some major environmental change occurs. In the environmental conditions that exist in Australia, a rainstorm is a major environmental change. The altered population that results following rain has reduced diversity. Localized environmental factors in a soil have minimal influence on the actinomycete population, but plants do modify the populations of the rhizosphere areas in association with their roots, the influence varying from species to species of plant. The sum total of the environmental factors that are involved in a change in geographic location has the greatest influence in inducing changes in the soil streptomycete populations. The varied populations do not remain constant during the year, but undergo significant seasonal changes. There is considerable stability in the actinomycete populations, however, with some populations showing little change for time periods exceeding a month, unless a major environmental change occurs.

3.3.1. General Ecological Aspects

Diversity indices can also be used to follow successional changes in a community. Theoretically, the diversity should increase from the low diversity of the pioneer populations to the higher and stable diversity of the climax community. This increase in diversity has been observed by Jordan and Staley (1976) for the periphyton community of Lake Washington, using submerged grids and electron-microscopic observations. The diversity, measured with the Shannon index, increased during a 10-day period; H' increased from 3.1 at day 1 to 4.2 at day 3, continuing to rise until it reached a maximum value of 4.8 on day 10 (Fig. 5). Bacteria were the dominant component during the pioneer stage of succession. During the successional process, some pioneer populations disappeared and the relative proportions of biomass shifted from heterotrophic eubacteria to algae and blue-green algae (cyanobacteria). In performing this study, Jordan and Staley took advantage of the distinctive morphologies of microorganisms in the periphyton community, which allowed them to distinguish taxonomic groups. They also considered the problems inherent in the use of the Shannon index to describe the diversity of the bacterial community. Two assumptions were necessary to justify use of the Shannon formula for determining the diversity index. One was that the distribution of organisms on the grid surface was random and not clumped or patchy. At the time of insertion of the grids into the habitat, one would expect that attachment to the grids would be random unless large clumps of detritus carrying bacteria were attach-

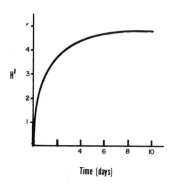

Figure 5. Successional changes in the diversity of a planktonic community. After Jordan and Staley (1976).

ing to the grids, but there was little evidence of this when the grids were examined following 24-hr incubation. After a period of incubation, one might expect that patchiness would occur due to reproduction of attached organisms resulting in the formation of microcolonies. Such was the case, but this patchiness was taken into account by determining the diversity index for clones (including single-cell microcolonies), rather than for individuals. The other assumption was that the total number of species on the grid was known at each time interval, but the total number of species attached to the grids is impossible to determine. Thus, the absolute values for the diversity indices determined in this study may be in error. Though this may be true, it is doubtful that the values determined relative to one another would be affected. For example, if there were twice as many species at each interval as were actually found in the 100 microcolonies counted and Basharin's estimator formula (Basharin, 1959) $E(H') - H' - (s - 1)/2N$ were applied, the new estimates of diversity for each time interval would be 2.97, 3.90, 4.03, and 4.46 for the 1-, 3-, 6-, and 10-day intervals, respectively. Therefore, the pattern of values relative to one another would not be affected.

Few studies on bacterial diversity can rely on morphological observations as the primary criteria for distinguishing taxa. Studies on diversity of bacterial communities, therefore, have most often employed the techniques of numerical taxonomy to define operational taxonomic units as defined by Sneath and Sokal (1973). The taxon as defined following clustering in numerical taxonomy is functionally equivalent to a species and can be used as such in the calculation of the diversity index. The use of microtiter methods and automated operations permits the gathering of sufficient data to characterize large numbers of isolates. The use of such techniques, however, like that of many others in microbial ecology, has the drawback of requiring the isolation of living strains for study and identification. Since any culture technique is selective, the use of viable isolates obtained from agar plates or other isolation sources introduces a bias into the procedure, and H' determined in this way underestimates total diversity of the bacterial community; if isolates are appropriately chosen at random, however, H' can estimate the diversity of that portion of the bacterial community that is capable of growth under the conditions defined by the isolation procedure. H' determined in this way is a measure of taxonomic diversity of certain populations, e.g., dominant aerobic heterotrophs.

Kaneko et al. (1977) employed this approach in a study of bacterial community diversity in nearshore regions of the Beaufort Sea. They found that there was an inverse relationship between bacterial biomass and species diversity in surface waters, but diversity was near maximal in sediments regardless of the size of the bacterial community. The inverse relationship between cell number and species diversity is in accordance with the general ecological principle that suggests that elevated populations in resource-limited habitats reflect the success of relatively few well-adapted (opportunistic) species. The similar levels of diversity in sediments at all locations suggest that these marine sediment communities are not severely stressed by environmental factors and thus achieve a high level of diversity characteristic of biologically accommodated communities. This study also showed that there were regular seasonal fluctuations in the species composition of the surface-water bacterial community. During summer, when such arctic regions experience continuous sunlight, the dominant bacterial populations found in surface waters are pigmented; during winter, when no direct sunlight reaches these waters, most bacteria are nonpigmented. Continuation of this study beyond the time period reported by Kaneko and colleagues (unpublished data) shows that the same levels of diversity are achieved at the same locations at the same times of year, but the species occurring in the community are not always the same. These data support the hypothesis that each year following the annual subice diatom bloom and breakup of coastal ice, a colonization process of surface waters occurs until the niches are filled; when the niches are filled, the community reaches a stable level of species diversity; the composition of the community depends on the opportunistic invasion of the surface waters by bacterial populations adapted to filling the available niches.

In a subsequent study, Hauxhurst et al. (1981) examined the characteristics of bacterial communities in the Gulf of Alaska. Taxonomic diversity was assessed using the Shannon–Weaver (H') and equitability (J') indices. Gulf of Alaska bacterial communities were characteristically diverse. Compared to arctic surface waters, the bacterial communities had higher diversities in surface waters in the Gulf of Alaska. This finding is in agreement with the ecological principle that fewer species are adapted to polar conditions and that species diversity decreases as one moves from equatorial to polar ecosystems. The diversity of sediment bacterial communities was not different in subarctic and arctic marine ecosystems, suggesting that these ecosystems are relatively unstressed and species diversity therefore does not decline as one moves toward the pole.

Using a similar approach, Martin (1980) studied the development of natural phytoplankton and associated heterotrophic bacteria. The diversity of the heterotrophic bacterial community depended on whether the phytoplankton was in a good physiological state, which had a simple stimulative effect on the bacteria, or whether the phytoplankton was senescent, which led to a strong selection of certain bacterial types. Judging from its characteristics, the bacterial community, which makes its appearance at the time of phytoplanktonic decline, occurs at the beginning of succession of heterotrophic bacteria in this experimental ecosystem. The heterotrophic bacterial succession differs from that of the more precocious phytoplankton community, which appears as soon as the nutrient salts are added. The dynamic pattern of the bacterial communities may be identical with that of other groups of organisms and reflects an evolution of poorly diversified, but highly active, pioneer stages to more organized stages in which activity may be reduced.

In the studies cited above, a large number of features were determined for each bacterial isolate. Acquiring the data base used for these taxonomic and subsequent community structure analyses was thus costly, especially in terms of the effort needed to generate and analyze the large amount of data. Griffith and Lovitt (1980) proposed simplifying the procedure by generating numerical profiles for studying bacterial diversity. They examined a total of 308 bacteria isolated from oil storage tanks. A numerical scoring method differentiated among the isolates and was used to estimate the diversity of the bacterial communities. Although the scoring methods suggested a higher diversity than did conventional identification, there was some consistency in the results produced by the two approaches. These workers concluded that scoring only nine or ten features, rather than the hundreds that are often used, could be useful for estimating and comparing diversity in other habitats.

The ability to calculate meaningful diversity indices based on limited numbers of features and limited numbers of representative isolates has been examined by Bianchi and Bianchi (1982), who found that a minimum of 30 features is necessary for differentiating bacterial taxa in marine samples. In their study on statistical sampling of bacterial strains for bacterial measurement, they further found that studying 20–30 strains is sufficient to determine the structure of the bacterial community based on the Shannon diversity index. This was determined by increasing the number of strains used to calculate H' until an asymptotic value for diversity was achieved (Fig. 6). A larger number of strains (>60) had to be examined to obtain a good estimate of species richness.

Although concern must be given to the interpretation of studies in which bacterial taxa are defined on the basis of a limited number of features, several such studies have yielded meaningful results. For example, Mills and Wassel (1980) used a limited number of features for calculating diversity of microbial

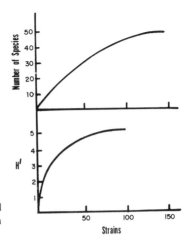

Figure 6. Effect of increasing number of bacterial strains identified on species richness and the Shannon diversity index. After Bianchi and Bianchi (1982).

communities collected from lake water and sediment samples in a study on the effects of acid mine drainage. Isolates were clustered by a numerical taxonomy approach in which a limited number (<20) of tests were used so that the groups obtained represented a level of resolution other than species. The numerical value of diversity for each sample was affected by the number of tests used; the relative diversity compared among several sampling locations was the same whether 11 or 19 characters were examined. In these studies, both the Shannon index (H') and rarefaction [$E(S)$] methods of estimating diversity were used. The number of isolates, i.e., sample size, strongly influenced the value of H', so that samples of unequal sizes could not be compared. As discussed in Section 2.1, rarefaction accounts for differences in sample size inherently so that such comparisons are made simple. Due to the type of sampling conducted by microbiologists, H' is estimated and not determined and requires a statement of error associated with it, and failure to report error provides potentially misleading results. Calculation of the variance of H' is not a simple matter and may be impossible when handling a large number of samples. With rarefaction, the variance of $E(S)$ is readily determined, facilitating the comparison of many samples. In their study, Mills and Wassel were able to show a decrease in species diversity of bacterial communities due to environmental stress.

3.3.2. Effects of Environmental Stress on Community Diversity

The effects of environmental stress on community structure have also been examined by Larrick *et al.* (1981), who studied the functional and structural changes of aquatic heterotrophic bacteria in response to thermal, heavy, and fly ash effluents. Heterotrophic bacterial populations were sampled at nine sites around a fossil-fuel power plant in Virginia to assess the ecological impact of the resulting effluents on naturally occurring heterotrophic microbes. The total colony-forming units (CFU) remained relatively high at all stations, ranging from 13,804 CFU/ml in the heavy-ash basin to 2630 CFU/ml in an un-influenced upstream station. The percentages of the total colony counts that were chromogenic were correlated with the physicochemical stresses and varied from a high of 59.0% in the reference New River station to 13.2% in the heavy-ash basin. A sequential comparison procedure produced diversity indices that ranged from 8.21 in upstream New River to 6.23 in the ash influence at the Adair Run station. The structure and function of bacterial communities in the ash basins were significantly different from the same parameters of populations inhabiting reference environments.

The influence of pollutants on bacterial community structure has also been reported by Peele *et al.* (1981), who studied the ecological impact of offshore dumping of pharmaceutical wastes. Normally, marine waters are dominated by gram-negative bacteria, e.g., *Vibrio* and *Pseudomonas* species. In the area of pharmaceutical dumping, the relative abundances of gram-negative species declined as gram-positive bacterial species increased in importance. *Staphylococcus, Micrococcus,* and *Bacillus* species were commonly found near the dump site, although these genera are rarely isolated from marine surface waters. The conclusion of this study was that oceanic dumping of pharmaceutical wastes alters the taxonomic composition of the bacterial community in the region of the dump site. The impact of marine pollution on community structure has also been examined by Horowitz and Atlas (unpublished data), who found that species diversity decreases when there is an input of petroleum hydrocarbons (Fig. 7). The greatest upset of normal community structure occurs with refined oils that have high concentrations of toxic hydrocarbons.

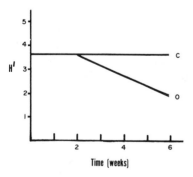

Figure 7. Effect of petroleum hydrocarbons on diversity of an arctic marine bacterial community. The diversity of the community exposed to oil (O) declines relative to controls (C) following an initial lag period. Based on Horowitz and Atlas (unpublished data).

4. Physiological Diversity

While taxonomic-diversity indices provide a useful descriptor of community structure, they do not provide information describing the cause of a particular state of heterogeneity within the species composition of a community. It is often of interest to understand the factors, especially environmental parameters and physiological functions of microbial populations, that interact to determine community structure. To this end, additional methods of assessing community diversity, based on the physiological properties of the member populations, have been developed.

Troussellier and Legendre (1981) have described an index of functional evenness for the study of bacterial communities, where the species level is poorly defined. This index bypasses the step of species identification, using directly as data a set of binary biochemical (and other) descriptors characterizing the microbial isolates. These workers have shown that this new index has the properties of the usual measure of evenness (J). The evenness of a bacterial community is defined as $E = I/I_{max} = 1/c$ log $0.5[\Sigma p_i$ log $p_i + (1 - p_i)$ log $(1 - p_i)]$ where p_i is a probability for the ith test being positive and c is the number of tests (descriptors) used. The functional evenness index (E) is based directly on the characteristics of the bacteria and thus directly measures the evenness of distribution of physiological attributes within the community. E is simpler to calculate than J, and additionally, values of E for different communities can be compared using statistical analyses.

Martin and Bianchi (1980) discussed the variations of structure (probable taxonomic generic groups, ecological profiles), diversity (Shannon index), and average catabolic potentialities [strain's average exoenzyme production (EAI) and average carbonaceous compound utilization (UAI)] of bacterial populations during two experimental phytoplankton blooms. They found that oligotrophic conditions are characterized by high diversity levels (H' from 3.60 to 4) and moderate catabolic potentialities (EAI and UAI close to 40%). During phytoplankton exponential growth, bacteria show an EAI stability, but there is an increase of UAI with maximal values at the beginning of chlorophyll plateau (52–57%) at higher values of diversity ($H' > 4$). Phytoplankton mortalities appear to cause an EAI increase and a decrease of both UAI and H' (1.50–2). Dominance by vibriolike organisms seems to be closely related to this period. Despite these similar patterns, many differences appear, from a taxonomic point of view, between experiments conducted seasonally; the autumn population is more diversified than the spring one. The use of these nutritional measures, in addition to taxonomic-diversity measurements, appears to be useful for spatial or temporal comparisons of community structure and function.

A similar approach was employed by Hauxhurst *et al.* (1981), who extended the principle of measuring nutritional versatilities to describe an index of physiological tolerance, i.e., of physiological diversity or heterogeneity, to describe community structure. Such analyses have provided evidence of annual variations in the member populations of a community based on the nutritional capabilities of the population, i.e., the enzymatic capability of utilizing particular substrates, and the availability of particular substrates in that habitat. The bacterial populations in these marine ecosystems were generally eurytolerant and nutritionally versatile. The maintenance of a high degree of informational heterogeneity was characteristic of these communities. It appears to be of adaptive advantage to maintain diverse populations with physiological tolerances the ranges of which exceed those experienced within the natural habitat and for the bacterial communities to possess a high degree of nutritional versatility within these marine ecosystems. As the source of food changes, so does the structure of the community. Analysis of physiological diversity among members of marine microbial communities also suggests that oligotrophic populations are physiologically and nutritionally diverse, capable of utilizing many substrates under varied environmental conditions, whereas copiotrophic populations are more specialized (less diverse), capable of extensive growth only under restricted conditions (Table II) (Horowitz *et al.*, 1983). Such extensions of conventional descriptors of community diversity provide useful additional information concerning the function, as well as the structure, of the microbial community.

Even greater information about the driving forces that determine community structure can be obtained using factor and principal-component analysis; the applicability of these techniques to studies in microbial ecology have been described in a review in Volume 2 of *Advances in Microbial Ecology* (Rosswall and Kvillner, 1978). Factor analysis has been used by Sundman (1970, 1973) to describe soil microbial populations, by Vaatanen (1980) to

Table II. Taxonomic, Physiological, and Nutritional Diversities of Oligotrophic and Copiotrophic Bacterial Populations from Subarctic Marine Waters

Bacterial population[b]	Diversity indices[a]								
	H'	P_H	P_T	P_S	N_c	N_a	N_{ca}	N_{aa}	N_h
Copiotrophs	4.1	0.60	0.80	0.60	0.39	0.33	0.35	0.46	0.16
Oligotrophs	2.8	0.80	0.80	0.81	0.35	0.59	0.71	0.72	0.50

[a]Indices: (H') Shannon diversity index; (P_x) physiological tolerance indices on a scale of 0–1; values near 0 indicate stenotolerance, i.e., a lack of physiological diversity: (P_H) tolerance index for pH; (P_T) tolerance index for temperature; (P_S) tolerance index for salinity; (N_x) nutritional versatility indices on a scale of 0–1; values near 0 indicate fastidiousness, i.e., a lack of nutritional diversity: (N_c) nutritional versatility for utilization of carbohydrates; (N_a) nutritional versatility for utilization of alcohols; (N_{ca}) nutritional versatility for utilization of carboxylic acids; (N_{aa}) nutritional versatility for utilization of amino acids; (N_h) nutritional versatility for utilization of hydrocarbons.
[b]Oligotrophs were isolated on a low-nutrient medium, copiotrophs on a peptone-based medium.

assess the impact of environmental factors on marine microbial communities, and by Toerien and co-workers (Toerien, 1970; Toerien *et al.,* 1969).

Holder-Franklin and co-workers have made extensive use of factor analysis to describe the diversity of freshwater microbial communities (Bell *et al.,* 1980, 1982, 1983; Holder-Franklin, 1981; Holder-Franklin *et al.,* 1978, 1981; Holder-Franklin and Wuest, 1982). Using factor analysis, they have been able to describe not only the community structure but also the factors, both environmental and physiological, that interact to establish a particular assemblage of microbial populations within the bacterial community of two Canadian rivers. In these studies, the factors controlling diurnal as well as seasonal variations were considered. These studies showed that seasonal and diurnal population shifts within the bacterial community are controlled in low-nutrient freshwater rivers primarily by temperature, nutritional versatility, dissolved oxygen, and conductivity and, less importantly, by pH and algal activity. There emerged from these investigations four factors that profiled the bacterial community and were common to both rivers. These factors were, in order of decreasing importance, fermentative metabolism, inorganic nitrogen metabolism, fluorescence–oxidative metabolism, and lack of starch hydrolysis. Several factors produced significant correlations with a range of physicochemical parameters. The correlation suggested an intricate algal–bacterial interaction. The oxidative-metabolism factor correlated with rainfall in one river, suggesting that the oxidative bacteria may be washed into the river from the surrounding land. In the other river, the oxidative–fermentative factor correlated negatively with sunshine. Factor analysis appears to be the most effective method for revealing correlations between bacterial characteristics and environmental parameters.

5. Genetic Diversity

Ultimately, diversity reflects the genetic composition of the species occurring together within the community. Microbiologists have begun to investigate the genetic composition of communities with the aim of determining the true basis of heterogeneity. The degree of genetic heterogeneity within the community should relate to the ability of the community to respond to environmental fluctuations and thus to the stability of the community. Communities with low genetic diversity should be less able to withstand severe environmental stress than those communities in which the genetic heterogeneity provides the elasticity needed for dealing with environmental modifications. How much energy a community can expend in maintaining a high level of genetic diversity depends on the energy resources available to the community; there is little question that the replication of extra segments of DNA needed to maintain genetic diversity is costly to the community in terms of energy resources.

Nevertheless, we see at the population level the evolution of mechanisms for ensuring the heterogeneity of the gene pool. Recombinational processes prevent the development of homogeneous DNA within a population, and even bacteria retain genetic exchange mechanisms for this purpose.

The recognition of genetic diversity within populations conflicts with some of the basic assumptions inherent in calculating species diversity, namely, that all individuals of the same species are identical and that all pairs of differing species are equally different (Hendrickson and Ehrlich, 1971). Clearly, at least the first assumption is in error for microbial populations. Genetic variability has been examined in many taxonomic studies aimed at determining genetic relatedness (Trüper and Kramer, 1981). For example, genetic diversity of *Rhizobium* species has been studied by Crow *et al.* (1981) and by Noel and Brill (1980). The genetic variability of naturally occurring bacterial populations was examined by Holder-Franklin *et al.* (1981), who examined the genetic diversity of bacterial taxa using the technique of DNA homology. They found that phenetic clusters are not genetically homologous. Nevertheless, they found that genetic variability within populations did not negate community structure analyses.

In addition to variability within the chromosomal DNA, the presence of plasmids within bacterial populations adds to the level of genetic variability. Plasmids provide a means of establishing diversity within populations of the same species (Starr and Schmidt, 1981). The distribution of plasmids within a community is thus a measure of the diversity of the populations in that community. Several groups, e.g., John and McNeill (1980), have examined plasmid diversity in clinical isolates; the presence of plasmids among clinical isolates is of extreme importance because of their role in determining resistance to antibiotics. Hada and Sizemore (1981) have studied the incidence of plasmids in marine *Vibrio* spp. isolated from an oil field in the northwestern Gulf of Mexico. They found that the oil field had noticeably more plasmid-containing strains (35 vs. 23% in the control site) and a greater number of plasmids per plasmid-containing strain (an average of 2.5 plasmids vs. 1.5 in the control site). Oil-field discharges might have resulted in increased plasmid incidence and diversity. Thus, under environmental stress, there appears to be an increase in plasmid richness and a decline in the evenness of plasmid distribution within the community. This study indicates that pollution stress alters the genetic diversity of the community. In a similar study on the role of plasmids in mercury transformation by bacteria isolated from the aquatic environment, Olson *et al.* (1979) found that genes conferring resistance to heavy metals occur with surprisingly high frequency within bacterial communities in the natural environment. They postulated that plasmids may provide for aquatic bacteria a fundamental and highly significant capacity for rapid response to introduction of allochthonous substrates in which genetic versatility and rapid response to environmental alteration can dictate survival.

One of the more interesting approaches for directly assessing the genetic diversity of a community has been described by Torsvik (1980). In her examination of a soil bacterial community, Torsvik extracted the DNA from the total assemblage of microorganisms. She then melted the DNA and allowed it to reanneal by lowering the temperature very slowly. Torsvik proposes that the slope of the reannealing curve is a measure of the heterogeneity of the DNA of the totality of the microbial community: a high slope for the reannealing curve indicates a high degree of DNA homology, as would occur if there was a dominant population of limited genetic heterogeneity—such a community would have low diversity; a low slope for the reassociation of the strands of DNA would indicate a high state of genetic heterogeneity, which would be characteristic of a diverse microbial community. Thus, by measuring the rate at which DNA reanneals, one can assay directly the state of community genetic diversity. The use of genetic techniques in microbial ecology should provide an improved basis for understanding community structure, the true state of diversity within microbial communities, and the degree of heterogeneity that must be stored within the microbial community to maintain stability.

6. Concluding Remarks

The measurement of microbial diversity provides insight into the ecological functioning of the community. Diversity measurements describe the heterogeneity of information within the community, in terms of both the total amount of information, which can be measured as species number, metabolic capabilities, or genetic potentialities, and the way in which the information is apportioned within the community. Although the specific cause-and-effect relationships are far from obvious, diversity is clearly related to stability. Communities, including microbial communities, evolve through a series of successional stages until a stable community structure is achieved at a certain level of diversity. Microorganisms, like all biological systems, respond to their biological and abiotic environment. The interactions of populations living together in a habitat establish a community structure in which the functional niches of the ecosystem are filled by member populations of the community. The diversity of a community reflects the status of population interrelationships within the community in dynamic terms. Changes in the environment lead to changes in the community structure, and as the community changes in terms of productivity and maturity, so does the diversity of the community.

There are several indices available for describing the state of ecological diversity. Species-diversity indices are not without problems in terms of when a particular index is appropriate to use, how to interpret a particular state of measured diversity, and particularly the ecological meaning of why a community has achieved a particular state of diversity. This being so, it is easy to misuse diversity measurements, and ecologists have too often fallen into the pitfalls of misinterpreting the meaning of diversity. A particular difficulty in the measurement of microbial diversity is the need to identify microbial species without bias; this can be accomplished by phycologists, but bacteriologists, relying on cultural methods to identify species, face the problem of selectivity and thus the inevitable underestimation of community diversity. Despite these problems, the diversity index, as a community parameter, can provide useful information in describing microbial communities. Diversity does represent the totality of potential activities and interactions of populations living together within the community. By extending the conventional diversity indices into the areas of physiological and genetic diversity, the microbial ecologist gains useful data on the functional status of a community. Diversity changes in response to environmental stress, and diversity measurements are therefore useful for monitoring changes due to pollution as natural perturbations of an ecosystem. With prudent use, diversity measurements provide one more way for developing an understanding of the ecology of microorganisms.

References

Adams, J. E., and McCune, E. D., 1979, Application of the generalized jackknife to Shannon's measure of information used as an index of diversity, in: *Ecological Diversity* (J. F. Grassle, G. Pattil, W. Smith, and C. Tallie, eds.), pp. 117–132, Statistical Ecological Series, Vol. 6, International Cooperative Publishing House, Burtonsville, Maryland.

Anderson, J. T., Roff, J. C., and Gerrath, J., 1981, The diatoms and dinoflagellates of Hudson Bay Canada, *Can. J. Bot.* **59**:1793–1810.

Bartha, Z., and Hajdu, L., 1979, Phytoplankton community structure studies on Lake Velence, Hungary. 1. Diversity, *Acta. Bot. Acad. Sci. Hung.* **25**:187–222.

Basharin, G. P., 1959, On a statistical estimate for the entropy of a sequence of independent random variables, *Theory Probab. Its Appl. (USSR)* **4**:333–336.

Bell, C. R., Holder-Franklin, M. A., and Franklin, M., 1980, Heterotrophic bacteria in two Canadian rivers. I. Seasonal variations in the predominant bacterial populations, *Water Res.* **14**:449–460.

Bell, C. R., Holder-Franklin, M. A., and Franklin, M., 1982, Correlations between predominant heterotrophic bacteria and physiochemical water quality parameters in two Canadian rivers, *Appl. Environ. Microbiol.* **43**:269–283.

Bell, C. R., Holder-Franklin, M. A., and Franklin, M., 1982, Seasonal fluctuations in river bacteria as measured by multivariate statistical analysis of continuous cultures, *Can. J. Microbiol.* **28**(8):959-975.

Belser, L. W., 1979, Population ecology of nitrifying bacteria, in: *Annual Review of Microbiology* (M. P. Starr, ed.), pp. 309–334, Annual Reviews, Palo Alto, California.

Belser, L., and Schmidt, E. L., 1978, Diversity in the ammonia oxidizing nitrifier population of a soil, *Appl. Environ. Microbiol.* **36**:584–588.

Bianchi, M. A., and Bianchi, A. J., 1982, Statistical sampling of bacterial strains and its use in bacterial diversity measurement, *Microb. Ecol.* **8**:61–69.

Bogaczewicz-Adamczak, B., 1978, The structure and production of phytoplankton in Lake Tynwalk, *Arch. Hydrobiol. Suppl.* **51**:328–336.

Brawley, S. H., and Adey, W. H., 1981, The effect of micrograzers on algal community structure in a coral reef microcosm, *Mar. Biol. (Berlin)* **61**:167–178.

Brillouin, L., 1962, *Science and Information Theory,* Academic Press, New York.

Bruno, S. F., Staker, R. D., and Sharma, G. M., 1980, Dynamics of phytoplankton productivity in the Peconic Bay estuary, Long Island, New York, USA, *Estuarine Coastal Mar. Sci.* **10**:247–263.

Cairns, J. C., 1969, Factors affecting the number of species in freshwater protozoan communities, in: *The Structure and Function of Freshwater Microbial Communities* (J. C. Cairns, ed.), pp. 219–248, Virginia Polytechnic Institute and State University, Blacksburg.

Cairns, J. C., Kuhn, D. L., and Plafkin, J. L., 1979, Factors affecting the number of species in freshwater protozoan communities, in: *The Structure and Function of Freshwater Microbial Communities* (J. C. Cairns, ed.), pp. 219–248, Virginia Polytechnic Institute and State University, Blacksburg.

Crow, V. L., Jarvis, B. D. W., and Greenwood, R. M., 1981, DNA homologies among acid producing strains of *Rhizobium, Int. J. Syst. Bacteriol.* **31**:152–172.

Dennis, B., and Patil, G. P., 1977, The use of community diversity indices for monitoring trends in water pollution impacts, *Trop. Ecol.* **18**:36–51.

Eloranta, P., 1976, Species diversity in the phytoplankton of some Finnish lakes, *Ann. Bot. Fenn.* **13**:42–48.

Eloranta, P., 1980, Annual succession of phytoplankton in one heated pond in Central Finland, *Acta Hydrobiol.* **22**:421–438.

Eloranta, P., and Kettunen, R., 1979, Phytoplankton in a watercourse polluted by a sulfite cellulose factory, *Ann. Bot. Fenn.* **16**:338–350.

Eloranta, P., and Kunnas, S., 1979, The growth and species communities of the attached algae in a river system in Central Finland, *Arch. Hydrobiol.* **86**:27–44.

Fagerberg, W. R., and Arnott, H. J., 1979, Seasonal changes in structure of a submerged blue-green algal bacterial community from a geothermal hot spring, *J. Phycol.* **15**:445–452.

Fagerberg, W. R., and Arnott, H. J., 1981, The structure of a geothermal stream community, *Bot. Gaz.* **142**:408–414.

Foissner, W., 1980, Species richness and structure of the ciliate community in small water bodies of the Austrian Alps Hohe-Tauern, *Arch. Protistenkd.* **123**:99–126.

Gamble, T. N., Betlach, M., and Tiedje, J. M., 1977, Numerically dominant denitrifying bacteria from world soils, *Appl. Environ. Microbiol.* **33**:926–939.

Gause, G. F., 1934, *The Struggle for Existence,* Williams and Wilkins, Baltimore.

Griffith, A. J., and Lovitt, R., 1980, Use of numerical profiles for studying bacterial diversity, *Microb. Ecol.* **6**:35–44.

Grimes, J. A., St. Clair, L. L., and Rushforth, S. R., 1980, A comparison of epiphytic diatom assemblages on living and dead stems of the common grass *Phragmites australis, Great Basin Nat.* **40**:223–229.

Hada, H. S., and Sizemore, R. K., 1981, Incidence of plasmids in marine *Vibrio* spp. isolated from an oil field in the northwestern Gulf of Mexico, *Appl. Environ. Microbiol.* **41**:199–202.

Hairston, N. G., Allan, A. D., Colwell, R. R., Fustuyma, D. J., Howell, J., Lubin, M. D., Mathias, J., and Vandermeer, J. H., 1968, The relationship between species diversity and stability: An experimental approach with protozoa and bacteria, *Ecology* **49**:1091–1101.

Hamada, S. E., and Farrand, S. K., 1980, Diversity among B-6 strains of *Agrobacterium tumefaciens, J. Bacteriol.* **141**:1127–1133.

Hanson, R. S., 1980, Ecology and diversity of methylotrophic organisms, in: *Advances in Applied Microbiology* (D. Perlman, ed.), pp. 3–40, Academic Press, New York.

Hatano, H., and Watanabe, Y., 1981, Seasonal change of protozoa and micrometazoa in a small pond with leaf litter supply, *Hydrobiologia* **85**:161–174.

Hauxhurst, J. D., Kaneko, T., and Atlas, R. M., 1981, Characteristics of bacterial communities in the Gulf of Alaska, USA, *Microb. Ecol.* **7**:167–182.

Hendrickson, J. A., and Ehrlich, P. R., 1971, An expanded concept of species diversity, *Not. Nat. Acad. Nat. Sci. Philadelphia* **439**:1–6.

Henebry, M. S., and Cairns, J., 1980a, The effect of island size, distance and epicenter maturity on colonization in freshwater protozoan communities, *Am. Midl. Nat.* **104**:80–92.

Henebry, M. S., and Cairns, J., 1980b, The effect of source pool maturity on the process of island colonization: An experimental approach with protozoan communities, *Oikos* **35**:107–114.

Holder-Franklin, M. A., 1981, Methods of studying population shifts in aquatic bacteria in response to environmental change, Scientific Series No. 124 (D. Carlisle, ed.), Inland Waters Directorate, Water Quality Branch, Department of the Environment, Ottawa, Canada.

Holder-Franklin, M. A., and Wuest, L. J., 1983, Factor analysis as an analytical method in microbiology, in: *Mathematical Methods in Microbiology* (M. Bazin, ed.), Academic Press, London (in press).

Holder-Franklin, M. A., Franklin, M., Cashion, P., Cormier, C., and Wuest, L. J., 1978, Population shifts in heterotrophic bacteria in a tributary of the Saint John River as measured by taxometrics, in: *Microbial Ecology* (M. W. Loutit and J. A. R. Miles, eds.), pp. 44–50, Springer-Verlag, Berlin.

Holder-Franklin, M. A., Thorpe, A., and Cormier, C. J., 1981, Comparison of numerical taxonomy and DNA–DNA hybridization in diurnal studies of river bacteria, *Can. J. Microbiol.* **27**:1165–1184.

Horowitz, A., Krichevsky, M. I., and Atlas, R. M., 1983, Characteristics and diversity of subarctic marine oligotrophic, stenoheterotrophic, and euryheterotrophic bacterial populations, *Can. J. Microbiol.* **29**:527–535.

Hudon, C. and Bourget, E., 1981, Initial colonization of artificial substrate: Community development and structure studied by scanning electron microscopy, *Can. J. Fish Aquat. Sci.* **38**:1371–1384.

Hulburt, E. M., 1963, The diversity of phytoplanktonic populations in oceanic, coastal and estuarine regions, *J. Mar. Res.* **21**:81–93.

Hungate, R. E., 1975, The rumen microbial ecosystem, *Annu. Rev. Microbiol.* **29**:39–66.

Hurlbert, S. H., 1971, The nonconcept of species diversity: A critique and alternative parameters, *Ecology* **52**:577–586.

Hutcheson, K., 1970, A test for comparing diversities based on the Shannon formula, *J. Theor. Biol.* **29**:151–154.

Hutchinson, G. E., 1959, Homage to Santa Rosalia, or Why are there so many kinds of animals?, *Am. Nat.* **93**:145–159.

Hutchinson, G. E., 1961, The paradox of the plankton, *Am. Nat.* **95**:137–145.

John, J. F., Jr., and McNeill, W. F., 1980, Plasmid diversity in multi-resistant *Serratia marcescens* isolates from affiliated hospitals, in: *Current Chemotherapy and Infectious Disease* (J. D. Nelson and C. Grassi, eds.), pp. 728–730, American Society for Microbiology, Washington, D.C.

Jordan, T. L., and Staley, J. T., 1976, Electron microscopic study of succession in the periphyton community of Lake Washington, *Microb. Ecol.* **2**:241–276.

Kaneko, T., Atlas, R. M., and Krichevsky, M., 1977, Diversity of bacterial populations in the Beaufort Sea, *Nature (London)* **270**:596–599.

Kemp, W. M., and Mitsch, W. J., 1979, Turbulence and phytoplankton diversity: A general model of the paradox of plankton, *Ecol. Model* **7**:201–222.

Kuserk, F. T., 1980, The relationship between cellular slime molds and bacteria in forest soil, *Ecology* **61**:1474–1485.

Lakkis, S., and Novel-Lakkis, V., 1981, Composition, annual cycle, and species diversity of the phytoplankton in Lebanese coastal water, *J. Plankton Res.* **3**:123–136.

Lapointe, B. E., Niell, F. X., and Fuentes, J. M., 1981, Community structure, succession, and production of seaweeds associated with mussel rafts in the Ría-de-Arosa, northwestern Spain, *Mar. Ecol. Prog. Ser.* **5**:243–254.

Larrick, S. R., Clark, J. R., Cherry, D. S., and Cairns, J., Jr., 1981, Structural and functional changes of aquatic heterotrophic bacteria to thermal heavy and fly ash effluents, *Water Res.* **15**:875–880.

LaZerte, B. D., and Watson, S., 1981, The prediction of lacustrine phytoplankton diversity, *Can. J. Fish Aquat. Sci.* **38**:524–534.

Legendre, L., and Legendre, P., 1982, *Numerical Ecology*, Elsevier, Amsterdam.

Levine, N. D., Corliss, J. O., Cox, F. E. G., Deroux, G., Grain, J., Honigberg, B. M., Leedale, G. F., Loeblich, A. R., Lom, J., Lynn, D., Meringeld, E. G., Page, F. C., Poljansky, G., Sprague, V., Vavra, J., and Wallace, F. G., 1980, A newly revised classification of the Protozoa, *J. Protozool.* **27**:27–58.

Lloyd, M., Zar, J. H., and Karr, J. R., 1968, On the calculation of informational measures of diversity, *Am. Midl. Nat.* **79**:257–272.

Luckinbill, L. S., 1979, Regulation, stability, and diversity in a model experimental microcosm, *Ecology* **60**:1098–1102.

MacArthur, R. H., and Wilson, E. O., 1963, An equilibrium theory of insular zoogeography, *Evolution* **17**:373–387.

MacArthur, R. H., and Wilson, E. O., 1967, *The Theory of Island Biogeography*, Princeton University Press, Princeton, New Jersey.

Margalef, R., 1951, Diversity of species in natural communities, *Publ. Inst. Biol. Apl. (Barcelona)* **9**:5–28.

Margalef, R., 1958, Information theory in ecology, *Gen. Syst.* **3**:36–71.

Margalef, R., 1961, Communication of structure in planktonic populations, *Limnol. Oceanog.* **6**:124–128.

Margalef, R., 1963, On certain unifying principles in ecology, *Am. Nat.* **97**:357–374.

Margalef, R., 1967, Some concepts relative to the organization of plankton, *Oceanogr. Mar. Biol. Annu. Rev.* **5**:257–289.

Margalef, R., 1968, *Perspectives in Ecological Theory*, University of Chicago Press, Chicago.

Margalef, R., 1979, Diversity, in: *Monographs on Oceanographic Methodology* (A. Sournia, ed.), pp. 251–260, UNESCO, Paris.

Marshall, J. S., Parker, J. I., Mellinger, D. L., and Lawrence, S. G., 1981, An *in situ* study of cadmium and mercury stress in the plankton community of 382 experimental lakes in the area of Northwestern Ontario, Canada, *Can. J. Fish Aquat. Sci.* **38**:1209–1214.

Martin, Y. P., 1980, Ecological succession of bacterial communities during the development of an experimental marine phytoplankton ecosystem, *Oceanol. Acta* **3**:293–300.

Martin, Y. P., and Bianchi, M. A., 1980, Structure, diversity, and catabolic potentialities of aerobic heterotrophic bacterial populations associated with continuous cultures of natural marine phytoplankton, *Microb. Ecol.* **5**:265–280.

May, R. M., 1976, *Theoretical Ecology: Principles and Applications*, W. B. Saunders, Philadelphia.

McIntosh, P. R., 1967, An index of diversity and the relation of certain concepts to diversity, *Ecology* **47**:392–404.

Menhinick, E. F., 1964, Density, diversity, and energy flow of arthropods in the herb stratum of a *Sericea lespedoza* stand, *Diss. Abstr.* **24**:4881–4882.

Mills, A. L., and Wassel, R. A., 1980, Aspects of diversity measurement of microbial communities, *Appl. Environ. Microbiol.* **40**:578–586.

Moore, J. W., 1979, Factors influencing the diversity, species composition, and abundance of phytoplankton in 21 arctic and subarctic lakes, *Int. Rev. Gesamte Hydrobiol.* **64**:485–499.

Moore, J. W., 1981, Benthic algae in littoral and profundal areas of a deep subarctic lake, 1981, *Can. J. Bot.* **59**:1026–1033.

Nelson, J. H., Stoneburner, D. L., Evans, E. S., Jr., Pennington, N. E., and Meisch, M. V., 1976, Diatom diversity as a function of insecticidal treatment with a controlled release formulation of chloropyrifos, *Bull. Environ. Contam. Toxicol.* **15**:630–634.

Noel, K. D., and Brill, W. J., 1980, Diversity and dynamics of indigenous *Rhizobium japonicum* populations, *Appl. Environ. Microbiol.* **40**:931–938.

Odum, H. T., Cantlon, J. E., and Kornicker, L. S., 1960, An organizational postulate for the interpretation of species–individual distributions, species entropy, ecosystem evolution and the meaning of species–variety index, *Ecology* **41**:395–399.

Olson, B. H., Barkay, T., and Colwell, R. R., 1979, Role of plasmids in mercury transformation by bacteria isolated from the aquatic environment, *Appl. Environ. Microbiol.* **38**:478–485.

Patrick, R., 1949, A proposed biological measure of stream conditions, based on a survey of the Conestoga Basin, Lancaster County, Pennsylvania, *Proc. Acad. Nat. Sci. Philadelphia* **101**:277–341.

Patrick, R., 1963, The structure of diatom communities under varying ecological conditions, *Ann. N. Y. Acad. Sci.* **108**:359–365.

Patrick, R., 1967, The effect of invasion rate, species pool, and size of area on the structure of the diatom community, *Proc. Natl. Acad. Sci. U.S.A.* **58**:1335–1342.

Patrick, R., 1976, The formation and maintenance of benthic diatom communities, *Proc. Natl. Acad. Sci. U.S.A.* **120**:475–484.

Patrick, R., 1977, Ecology of freshwater diatom communities, in: *The Biology of Diatoms* (D. Werner, ed.), Botanical Monographs, Vol. 13, Blackwell, Oxford.

Patrick, R., Hohn, M. H., and Wallace, J. H., 1954, A new method for determining the pattern of diatom flora, *Not. Nat. Acad. Nat. Sci. Philadelphia* **259**:1–12.

Patten, B. C., 1961, Competitive exclusion, *Science* **134**:1599–1601.

Patten, B. C., 1963, Species diversity in net phytoplankton of Raritan Bay, *J. Mar. Res.* **20**:57–75.

Peele, E. R., Singleton, F. L., Deming, J. W., Cavari, B., and Colwell, R. R., 1981, Effects of pharmaceutical wastes on microbial populations in surface waters at the Puerto Rico dump site in the Atlantic Ocean, *Appl. Environ. Microbiol.* **41**:873–879.

Peet, R. K., 1974, The measurement of species diversity, *Annu. Rev. Ecol. Syst.* **5**:285–308.

Pielou, E. C., 1966a, Shannon's formula as a measure of species diversity: Its use and misuse, *Am. Nat.* **100**:463–465.

Pielou, E. C., 1966b, The measurement of diversity in different types of biological collections, *J. Theor. Biol.* **13**:131–144.

Pielou, E. C., 1969, *An Introduction to Mathematical Ecology*, Wiley-Interscience, New York.

Pielou, E. C., 1975, *Ecological Diversity*, Wiley-Interscience, New York.

Pielou, E. C., 1977, *Mathematical Ecology*, Wiley-Interscience, New York.

Pollingher, U., 1981, The structure and dynamics of the phytoplankton assemblages in Lake Kinneret, Israel, *J. Plankton Res.* **3**:93–106.

Preston, F. W., 1962a, The canonical distribution of commonness and rarity. I, *Ecology* **43**:185–215.

Preston, F. W., 1962b, The canonical distribution of commonness and rarity. II, *Ecology* **43**:410–432.

Rees, J. T., 1979, Community development in fresh water microcosms, *Hydrobiologia* **63**:113–128.

Revelante, N., and Gilmartin, M., 1980, Microplankton diversity indices as indicators of eutrophication in the northern Adriatic Sea, *Hydrobiologia* **70**:277–286.

Richerson, P., Armstrong, R., and Goldman, C. R., 1970, Contemporaneous disequilibrium, a new hypothesis to explain the "paradox of the plankton," *Proc. Natl. Acad. Sci. U.S.A.* **67:**1710–1714.

Rosswall, T., and Kvillner, E., 1978, Principal-components and factor analysis for the description of microbial populations, in: *Advances in Microbial Ecology,* Vol. 2 (M. Alexander, ed.), pp. 1–48, Plenum Press, New York.

Saifullah, S. M., and Moazzam, M., 1978, Species composition and seasonal occurrence of centric diatoms in a polluted marine environment, *Pak. J. Bot.* **10:**53–64.

Samuels, W. B., Uzzo, A., and Nuzzi, R., 1979, Correlations of phytoplankton standing crop, species diversity and dominance with physical–chemical data in a coastal salt pond, *Hydrobiologia* **64:**233–238.

Sanders, H. L., 1968, Marine benthic diversity: A comparative study, *Am. Nat.* **102:**243–282.

Sepers, A. B. J., 1981, Diversity of ammonifying bacteria, *Hydrobiologia* **83:**343–350.

Shannon, C. E., 1948, A mathematical theory of communication, *Bell Syst. Technol.* **27:**379–423.

Shannon, C. E., and Weaver, W., 1949, *The Mathematical Theory of Communications,* University of Illinois Press, Urbana.

Sheldon, A. L., 1969, Equitability indices: Dependence on species count, *Ecology* **50:**466–467.

Simberloff, D., 1972, Properties of the rarefaction diversity measurement, *Am. Nat.* **106:**414–418.

Simberloff, D., 1978, Use of rarefactions and related methods in ecology, in: *Biological Data in Water Pollution Assessment: Quantitative and Statistical Analyses* (K. L. Dickson, J. Cairns, Jr., and R. J. Livingston, eds.), pp. 150–165, ASTM STP 652, American Society for Testing Materials, Philadelphia.

Simpson, E. H., 1949, Measurement of diversity, *Nature (London)* **163:**688.

Slater, J. H., 1978, The role of microbial communities in the natural environment, in: *The Oil Industry and Microbial Ecosystems* (K. W. A. Chater and H. J. Somerville, eds.), pp. 137–154, Heyden and Son, London.

Slater, J. H., 1980, Physiological and genetic implications of mixed population and microbial community growth, in: *Microbiology—1980* (D. Schlessinger, ed.), pp. 314–316, American Society for Microbiology, Washington, D.C.

Sneath, P. H. A., and Sokal, R. R., 1973, *Numerical Taxonomy—The Principles and Practice of Numerical Classification,* W. H. Freeman, San Francisco.

Sousa, W. P., 1979a, Experimental investigations of disturbance and ecological succession in a rocky intertidal algal community, *Ecol. Monogr.* **49:**227–254.

Sousa, W. P., 1979b, Disturbance in marine intertidal boulder fields: The nonequilibrium maintenance of species diversity, *Ecology* **60:**1225–1239.

Staley, J. T., 1980, Diversity of aquatic heterotrophic bacterial communities, in: *Microbiology—1980* (D. Schlessinger, ed.), pp. 321–322, American Society for Microbiology, Washington, D.C.

Stanley, P. M., and Schmidt, E. L., 1981, Serological diversity of *Nitrobacter* spp. from soil and aquatic habitats, *Appl. Environ. Microbiol.* **41:**1069–1071.

Starr, M., and Schmidt, J. M., 1981, Prokaryote diversity, in: *The Prokaryotes* (M. P. Starr, H. Stolp, H. G. Trüper, A. Balows, and H. G. Schlegel, eds.), pp. 3–42, Springer-Verlag, Berlin.

Starr, M., and Skerman, V. D. B., 1965, Bacterial diversity: The natural history of selected morphologically unusual bacteria, *Annu. Rev. Microbiol.* **19:**407–454.

Stevenson, R. J., 1981, Microphytobenthos accumulation and current, Doctoral dissertation, 172 pp., School of Natural Resources, University of Michigan, Ann Arbor.

Stevenson, R. J., 1983, How currents on different sides of substrates in streams affect mechanisms of benthic algal accumulation, *Int. Rev. Gesamte Hydrobiol.* (in press).

Stevenson, R. J.,and Stoermer, E. G., 1981, Quantitative differences between benthic algal communities along a depth gradient in Lake Michigan, *J. Phycol.* **17**:29–36.

Stevenson, R. J., and Stoermer, E. F., 1982, Abundance patterns of diatoms on *Cladophora* in Lake Huron with respect to a point source of wastewater treatment plant effluent, *J. Great Lakes Res.* **8**:184–195.

Stockner, J. G., 1968, The ecology of a diatom community in a thermal stream, *Br. Phycol. Bull.* **3**:501–514.

Stockner, J. G., and Benson, W. W., 1967, The succession of diatom assemblages in the recent sediments of Lake Washington, *Limnol. Oceanogr.* **12**:513–532.

Sullivan, M. J., 1978, Diatom community structure: Taxonomic and statistical analyses of a Mississippi, USA, salt marsh, *J. Phycol.* **14**:468–475.

Sullivan, M. J., 1981, Effects of canopy removal and nitrogen enrichment on a *Distichlis spicata*–edaphic diatom complex, *Estuarine Coastal Shelf Sci.* **13**:119–130.

Sundman, V., 1970, Four bacterial populations characterized and compared by a factor analytical method, *Can. J. Microbiol.* **16**:455–464.

Sundman, V., 1973, Description and comparison of microbial populations in ecological studies with the aid of factor analysis, *Bull. Ecol. Res. Comm. (Stockholm)* **17**:135–141.

Swift, M. J., 1976, Species diversity and the structure of microbial communities in terrestrial habitats, in: *The Role of Terrestrial and Aquatic Organisms in Decomposition Processes* (J. M. Anderson and A. MacFadyen, eds.), pp. 185–222, Halsted Press, New York.

Sze, P., 1980, Seasonal succession of phytoplankton in Onondaga Lake, New York, *Phycologia* **19**:54–59.

Thompson, G. G., and Ho, J., 1981, Some effects of sewage discharge upon phytoplankton in Hong Kong, *Mar. Pollut. Bull.* **12**:168–173.

Tinnberg, L., 1979, Phytoplankton diversity in Lake Norrviken, Sweden, 1961–1975, *Holarct. Ecol.* **2**:150–159.

Tipper, J. C., 1979, Rarefaction and rarefiction—the use and abuse of a method in paleoecology, *Paleobiology* **5**:423–434.

Toerien, D. F., 1970, Population description of the non-methanogenic phase of anaerobic digestion. III. Non-hierarchical classification of isolates by principal component analysis, *Water Res.* **4**:305–314.

Toerien, D. F., Hattingh, W. H. J., Kotze, J. P., Thiel, P. G., and Sibert, M. L., 1969, Factor analysis as an aid in an ecological study of anaerobic digestion, *Water Res.* **3**:129–140.

Torsvik, V. L., 1980, DNA from soil bacteria, Presented at the Second International Symposium on Microbial Ecology, University of Warwick.

Trousellier, M., and Legendre, P., 1981, A functional evenness index for microbial ecology, *Microb. Ecol.* **7**:283–296.

Trüper, H. G., and Kramer, J., 1981, Principles of characterization and identification of prokaryotes, in: *The Prokaryotes* (M. P. Starr, H. Stolp, H. G. Trüper, A. Ballows, and H. G. Schlegel, eds.), pp. 176–193, Springer-Verlag, Berlin.

Vaatanen, P., 1980, Factor analysis of the impact of the environment on microbial communities in the Tvarminne Area, southern coast of Finland, *Appl. Environ. Microbiol.* **40**:55–61.

Walker, N., 1978, On the diversity of nitrifiers in nature, in: *Microbiology—1978* (D. Schlessinger, ed.), pp. 346–347, American Society for Microbiology, Washington, D.C.

Watling, L., Bottom, D., Pembroke, A., and Maurer, D., 1979, Seasonal variations in Delaware Bay, USA, phytoplankton community structure, *Mar. Biol.* **52**:207–216.

Whittaker, R. H., 1975, *Communities and Ecosystems*, Macmillan, New York.

Wilhm, J., Cooper, J., and Namminga, H., 1978, Species composition, diversity, biomass, and chlorophyll of periphyton in Greasy Creek, Red-Rock Creek, and the Arkansas River, Oklahoma, USA, *Hydrobiologia* **57**:17–24.

Wilson, C. J., and Holmes, R. W., 1981, The ecological importance of distinguishing between living and dead diatoms in estuarine sediments, *Br. Phycol. J.* **16**:345–349.

Woodwell, G. M., and Smith, H. H. (eds.), 1969, *Diversity and Stability in Ecological Systems,* Brookhaven Symposia in Biology, No. 22, Brookhaven National Laboratory, Upton, New York.

Yeh, W. K., and Ornston, L. N., 1980, Origins of metabolic diversity: Substitution of homologous sequences into genes for enzymes with different catalytic activities, *Proc. Natl. Acad. Sci. U.S.A.* **77**:5365–5369.

Yoshitake, S., 1981, Drifting algae in the river Tama, Japan: Dominant species, diversity index, and Motomura's index, *Jpn. J. Phycol.* **29**:117–120.

2

Microbial Ecology of Desert Soils

J. SKUJIŅŠ

1. Introduction

For many years, arid desert soils were considered economically unimportant, and any ecological research, including the examination of microbial characteristics, was sporadic. During the past two decades, however, the economic and agricultural utilization of arid lands has emerged as a critical element in maintaining and improving the world's food supply; consequently, biological and environmental research on these soils has increased. The expansion of deserts (desertification process) due to human impact, often in combination with adverse climatic disasters, has reinforced the necessity of understanding biological processes in xeric environments. In comparison with the body of knowledge about physical processes and about floral and faunal aspects in marginally utilizable arid areas, relatively little detailed information about arid-soil biological properties exists. Although certain patterns of arid-soil biological properties have emerged from the examination of several desert ecosystems, it is still premature to generalize about soil biological characteristics on a global desert biome level. Considerably more information is available on subhumid and semiarid cultivated soils (Focht and Martin, 1979).

Desert soils are usually characterized by high soil pH and often by high salinity, both of which influence the activities of soil microorganisms. Arid areas may include saline surface deposits and hypersaline bodies of water, for example, the Dead Sea in Israel and the Great Salt Lake in the Great Basin Desert, United States, inhabited by unique microbial populations. The ecology and physiology of halophilic organisms have been extensively studied (Brock, 1979; W. D. Williams, 1981), forming a part of "biosaline research" (Bishay

J. SKUJIŅŠ • Department of Biology and Ecology Centers, Utah State University, Logan, Utah 84322.

and McGinnies, 1979). Similarly, the biology of the unique cold deserts of Antarctica has been reviewed (Llano, 1972; Heinrich, 1976).

At the turn of the 20th century, most desert soils were considered abiotic and sterile. In 1912, Lipman (1912) first demonstrated that desert soils were inhabited by numerous microorganisms when he described microbiological characteristics of California deserts. Additional pioneering work was done in the Sahara. Rivkind (1929), for example, analyzed soils collected along the Alger-Le Hoggar route. The extensive research activities of Killian and Feher (1935, 1938, 1939) in the northern Sahara greatly facilitated our understanding of the microbial activities in desert soils. This body of research clearly demonstrated that arid soils contain a considerable number and variety of microorganisms, although often to a lesser extent than do soils of more mesic zones. In the United States during this period, work on arid-soil biology was sporadic. Oberholzer (1936) reported on the biological activities of Arizona desert soils. During the 1950s and 1960s, important research appeared about the desert soil microorganisms in Australia, the U.S.S.R., France, Egypt, and the United States. In Egypt, a group of investigators at the Ain Shams University, Cairo, investigated microbial flora of desert plant rhizospheres. During the 1960s, Cameron and co-workers examined numerous desert environments, and they published a series of reports about arid-soil microorganisms in the United States and Chile (Cameron and Blank, 1965; Cameron, 1969) and in Antarctica (Cameron *et al.,* 1976). Exhaustive dissertations by Dommergues (1962a,b) and by Sasson (1967) provided further insights into Saharan desert microbiology and into xeric microbial physiology. During the last two decades, numerous investigators in various parts of the world have contributed significantly to our understanding of the microbial ecology of desert soils. Sasson (1972) reviewed aspects of microbiology in arid environments, emphasizing the ecology of *Azotobacter.* Binet (1981) analyzed microbial effects on mineral cycling in arid ecosystems (information included up to 1975), and Friedmann and Galun (1974) published an excellent review about the biology of desert algae, lichens, and fungi.

2. Arid Environment

2.1. Definition of Desert

Desert ecosystems are characterized by lack of moisture, and the biological activities in the arid desert soils are regulated by ephemeral water availability. Various terms in the literature describe arid ecosystems such as "desert," "semidesert," "steppe," "subdesert," "semiarid," "arid grasslands," and others, depending on the vegetative cover and total precipitation. It has become evident, however, that for arid ecosystems, the interaction of precipitation, temperature, evaporation, and evapotranspiration is the defining factor for total water availability. Thornthwaite (1948) first developed a system for defining arid and semiarid environments. This system was recently reexamined and improved by Bailey (1979). In principle, these systems are based on the definition of water deficit in the environment, and in practice, they may be visualized as a relationship between mean annual precipitation and mean annual temperature (Fig. 1).A definition based on Thornthwaite's system, but including certain agroeconomic factors (Meigs, 1953), is also used for applied purposes.

Arid and semiarid ecosystems extend from the tropical to subalpine zones and into Antarctica, and from sea level to above 3000 m altitude. Due to the extreme variations in geographic and climatic patterns, the global arid environment (i.e., biome) consists of highly diversified and heterogeneous ecosystems. According to Meigs (1957), 19% (27.6 $\times$ 10^6 km^2) of the global land area is covered by arid deserts and 14.6% (21.2 $\times$ 10^6 km^2) by semiarid deserts, excluding Antarctica.

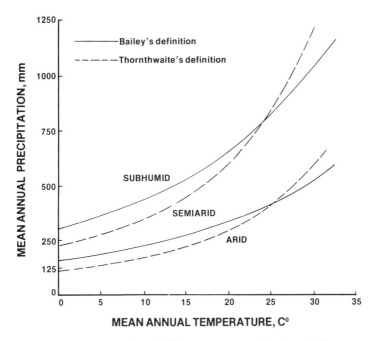

Figure 1. Definitions of arid and semiarid environments as functions of the mean annual temperature and precipitation. Adapted from Bailey (1979).

2.2. Soils

Soil microorganisms in arid environments are subjected to the same external physical factors as in other soils, except that the organisms are subjected to prolonged harsh environmental extremes. Frequently, it is difficult to pinpoint the individual influence of specific physical factors on organisms because the factors interact and produce a complex interrelated effect.

In addition to water availability, the great diversity of the ecological environment for microorganisms is due to the variability of soil types and their characteristics, including the quality and quantity of available organic matter. About 41% of desert soils belong to Entisols (soils without a characteristic profile development). Entisol areas are easily subjected to shifting dune formation (Psamments). Although the dune material is primarily sandy, it may also be gypsic (White Sands, New Mexico) or calcareous (Death Valley, California). Aridisols with a minimal profile development constitute 36% of the desert areas, and are typically present in the Great Basin and Sonoran deserts of the United States and in the Libyan Desert, adjoining the Entisols of Sahara proper. The surface of Aridisols may be subjected to localized shifting dune formation, especially when external forces destroy the surface cover. Other soils found in arid areas are Alfisols, Mollisols, and Vertisols, usually present in marginal desert semiarid areas.

A number of subsurface horizons, usually within 1 m depth, may have developed in soils that influence the water behavior and regime, such as argillic (clay accumulation), a fine-textured cambic, Na^+-rich natric, or other salt-accumulating salic horizons. Soil biological characteristics are severely affected by the often prevalent calcic ($CaCO_3$-enriched "petrocalcic" or "caliche"), gypsic, and duripan (silicic) horizons, which form hard, cemented layers, mostly impenetrable by plant roots and restrictive to water movement. Excellent reviews of desert soils have been published by Fuller (1974) and Dregne (1976). The latter publication contains informative listings of comparative international desert-soil classification nomenclature.

2.3. Water Factor

The effect of water as the ultimate limiting factor for microbial activities is more evident in desert soils than in other more mesic and humid environments. A large body of descriptive data has been gathered on microbial activities in environments with limited water availability, but the understanding of basic principles governing the physiology, intracellular processes, and adaptation of organisms to arid situations is meager.

For the description of microbial activities with respect to water availability, the total water content in soil is inconsequential; of far greater significance is the soil water potential, which essentially describes the force by which water is held in soil and may be related to the energy that microorganisms should expend for water assimilation. To assimilate water held in soil at -0.1 megapascal (MPa) (-1 bar), for example, organisms need approximately 2 joules (0.5 calorie) per mole (McLaren and Skujiņš, 1967). Cook and Papendick (1970), among other investigators, have ably demonstrated that microbial response to water content in different soils may be best compared in terms of soil water potentials rather than in terms of water content based on weight (Fig. 2).

The total soil water potential (ψ_w) is the sum of matric (ψ_m) and osmotic (ψ_s) potentials, expressed in energy units per mass (pascals) (-1 bar $= -100$ kPa $= -0.1$ MPa). Thermodynamically, the water potential may be expressed according to the equation

$$\psi_w = \frac{RT}{V_m} \ln \frac{p}{p_0}$$

where ψ_w is the water potential of the soil, R is the universal gas constant (8.2 MPa cm^3/mole K), T is the temperature (K), V_M is a molar volume of water (18 cm^3/mole), p is the vapor pressure of soil air, and p_0 is the vapor pressure of saturated air at the same temperature as the soil air. The ratio p/p_0 is also essentially the water activity, a_w, or the relative humidity of soil air, $a_w \times 100$. By substituting the respective values in the equation, one may write $\psi_w = 1.065$ $T \log a_w$, useful for practical applications (ψ_w in this equation is expressed in MPa). Griffin and Luard (1979) and Griffin (1981) perceptively described the thermodynamics of water relationships with respect to microbial ecology in arid soils. Practical approaches to the examination of soil water relationships, especially by utilizing thermocouple psychrometers, were delineated by Hanks and Ashcroft (1980).

3. Microbial Distribution

Reflecting the extensive heterogeneity of edaphic and climatic conditions in arid soils, the microbial numbers reported in the literature have considerable variation in range. Rougieux [(1966) (cited in Binet, 1981)] noted only 90 microorganisms present per gram of barren reg soil, whereas Hethener (1967) determined the presence of 80×10^6 microorganisms/g soil from *Cupressus dupreziana*-dominated soil in the Tassili N'Ajjer area of the Sahara. There is also similar variability in the reported fungal numbers. A major difficulty in interpreting the microbial numbers is due to the wide variety of media and

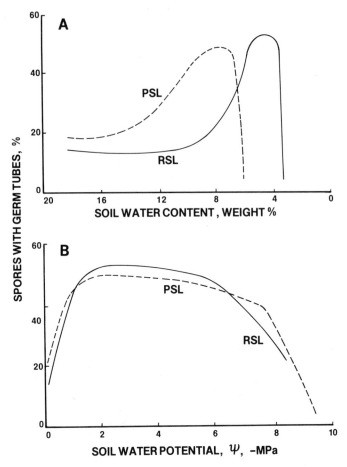

Figure 2. Comparison of the influence of soil water content (A) and of soil water potential (B) on spore germination of *Fusarium roseum* f. sp. *cerealis* "Culmorum" in Ritzville (RSL) and Palouse (PSL) silt loams, 72 hr following glucose and ammonium sulfate amendment. Adapted from Cook and Papendick (1970).

environmental conditions during incubation. Apparently, there are no studies in the literature in which authors based their experimental results on modern methods in desert-soil research, such as direct counting by fluorescence. In any case, it is evident that there are large numbers of various microorganisms present in arid soils. Because the bacterial numbers were estimated by plate counting, one can assume that the actual numbers are at least 50 times higher than estimated. Considering the low amount of organic matter in desert soils, it is apparent that microorganisms represent a considerably larger proportion of the total organic matter in desert soils than in other ecosystems.

3.1. Bacteria and Actinomycetes

The microbial community in desert soils is similar to that in soils of other ecosystems, and essentially all the functional and systematic groups may be found in arid soils. In diverse Saharan soils, Killian and Feher (1935, 1938, 1939) found bacteria belonging to 98 different species, the majority of which were spore-formers. The northern Egyptian desert (Omayed site) soils contained 0.08×10^6 to 0.81×10^6 bacteria/g soil and 7000–98,000 actinomycetes/g soil, whereas in the calcareous littoral dunes with sparse vegetation (Gharbaniat site), there were 0.01×10^6 to 2.2×10^6 bacteria/g and 2000–150,000 actinomycetes/g. In both sites, spore-formers constituted 6% of the total bacterial population (Abdel-Ghaffar and Fawaz, 1977). According to Mahmoud et al. (1964), Egyptian arid soils contained more than 32% spore-formers as compared to 8% in cultivated agricultural soils.

In the Tunisian Presaharan Desert, in the vicinity of Medenine, the aerobic microbial numbers were an order of magnitude less than those in the southwestern United States deserts; the mean numbers were 0.35×10^6 aerobic microorganisms/g and 0.09×10^6 actinomycetes/g (Johnson, 1973; Skujiņš, 1977b). Various coryneforms constituted the dominant organisms in these soils, followed by *Pseudomonas, Acinetobacter, Bacillus, Proteus,* and *Micrococcus*. Although *Bacillus* spp. are usually prevalent in desert soils, Johnson (1973) indicated that this genus was scarce in Tunisian Presaharan soils. In addition, he noted that the distribution of *M. roseus* is associated with ants, which act as carriers of this organism.

The mean values for three sites in the Chihuahuan Desert, New Mexico, were 1.5×10^6, 0.98×10^6, and 0.56×10^6 aerobic bacteria/g; the Sonoran Desert gave 1.7×10^6 and the Mojave Desert 0.83×10^6 aerobes/g. In the Great Basin Desert, the values ranged from 0.93×10^6 (Pine Valley) to 4.0×10^6 (Curlew Valley) aerobic bacteria/g. In Curlew Valley, the mean value for actinomycetes was 1.9×10^6/g, representing 47% of the estimated aerobic microbial population (Skujiņš and West, 1973; Skujiņš, 1977a). Actinomycetes of Arabian desert soils have been described by Elwan and Diab (1976) and Diab and Zaidan (1976).

The actinomycetes may contribute 50% of the total microbial bacterial population in desert soils. However, Hethener (1967) reported that actinomycetes represented only 1–2% of the population in the sandy soils of Tassili N'Ajjer, and Rougieux [(1966) (cited in Binet, 1981)] indicated that actinomycetes might not even be detectable in certain Saharan soils.

Myxobacters were isolated from arid soils in Monterrey, Mexico (Brockman, 1976), and in Arizona (Larkin and Dunigan, 1973; Reichenbach, 1970). The dominant genera were *Myxococcus, Sorangium, Cystobacter, Stigmatella, Polyangium,* and *Archangium*. The species diversity in arid soils decreased only slightly as compared to their presence in subhumid and mesic soils.

Representative reported numbers of microorganisms in desert soils are listed in Table I. It is not always possible to ascertain from the available literature either the climatic conditions at the time of soil collection or the media used.

3.2. Fungi

The fungal numbers in desert soils reported in the literature are propagule numbers apparently obtained by plate counting on various media. As far as can be ascertained from literature data, the hyphal length was not measured, and the actual biomass of fungi in these soils is unknown. Killian and Feher (1939) reported 1000–261,000 propagules/g soil in the Sahara Desert, listing 28 species with a predominance of *Penicillium, Syncephalastrum, Trichoderma,* and *Aspergillus.* The usefulness of these early studies, however, is questionable due to inexact taxonomy (Friedmann and Galun, 1974). Other representative values included 150,000–250,000 propagules/g soil from the Saharan Beni Ounif region (Pochon *et al.,* 1957), 50–34,000/g soil in the Sonoran Desert (Ranzoni, 1968), and between none and 12,000 organisms/g in the southwestern Asian deserts of the U.S.S.R. (Lobova, 1960). In the Presaharan Desert near Medenine, Tunisia, the fungal propagule numbers varied between 1000 and 130,000/g soil. The surface 1 cm of soil consistently had

Table I. Representative Microbial Numbers per Gram of Soil in Selected Desert Surface Soils

Soil	Aerobic bacteria and actinomycetes	Fungi	Reference
Kara-Kum Desert			
Crusty solonetzous soil	8.6×10^4	2.5×10^3	Lobova (1960)
Sandy gray-brown soil	6.7×10^5	1.3×10^3	Lobova (1960)
Sandy soil under vegetation	4.1×10^5	6.3×10^3	Lobova (1960)
Uzboi Desert, takyr	3.9×10^4	0	Lobova (1960)
Mojave Desert			
Nevada	2.9×10^6 to 1.6×10^7	3.6×10^3 to 5.0×10^4	Vollmer *et al.* (1977)
California	9×10^4	3.4×10^5	Cameron (1969)
Sahara Desert	2.0×10^3 to 1.5×10^6	—	Pochon *et al.* (1957)
Sahara Desert	1.0×10^6 to 8.0×10^6	—	Hethener (1967)
Sahara Desert (Abu Simbel)	1.6×10^5	15	Cameron (1969)
Atacama Desert	< 10	0	Cameron (1969)
Sonoran Desert			
Mammoth	1.5×10^6	170	Cameron (1969)
Baja California	2.5×10^6	4.9×10^3	Cameron (1969)
Chihuahuan Desert	1.7×10^5	2×10^3	Cameron (1969)

numbers of propagules one or two orders of magnitude higher than did the subsurface soils, where the propagule numbers averaged 1000/g soil. The dominant genus was *Alternaria* (Skujiņš, 1977b). The studies by Nicot (1955, 1960) of the Saharan soil with some vegetation indicated that *Alternaria, Curvularia, Helminthosporium,* and *Stemphylium* dominated the mycoflora. In addition, *Fusarium* and *Mucor* were present in the surface 5 cm of soil. The fungal diversity sharply decreased with depth. Below 10 cm, *Aspergillus fumigatus* among few other *Aspergillus* spp. and *Penicillium* spp. dominated. A sparse population, represented only by *Penicillium*, existed in the Grand Erg sand dunes. Similarly, Cameron et al. (1965) found only 1–2 *Penicillium* propagules/g soil in the extremely arid Atacama Desert near Antofagasta.

Naguib and Mouchacca (1971) examined the mycoflora of Egyptian desert soils from 31 sites. The number of propagules varied from 400 to 3200/ g in noncultivated soils. Saline soils and soil crusts contained between 1200 and 3400 propagules/g. The dominant species were *Mucor fragilis, Aspergillus fumigatus, A. niger, A. flavus, A. nidulans, A. ustus, A. terreus, Penicillium chrysogenum, P. notatum, P. cyclopium, P. funiculosum, Paecilomyces varioti, Fusarium moniliforme,* and Mycelia sterilia. A total of 106 fungal species were isolated. Rayss and Borut (1958) and Borut (1960) described an equally rich mycoflora in the northern Negev Desert, where they noted the presence of 79 fungal species, among them listing 21 *Penicillium* spp. and 14 *Aspergillus* spp., in addition to 18 Dematiaceae, 6 Mucedinaceae, 4 Sphaeropsidaceae, 2 Stilbaceae, 3 Tuberculaeraceae, 3 Mycelia sterilia, and 4 phycomycetes.

The mean values for six California desert profiles ranged from 1000 to 34,000 propagules/g soil (average 2500 propagules/g) (Cameron and Blank, 1965). In the Great Basin Desert (Curlew Valley), the numbers were between none (1 m depth) and 120,000/g (surface, rainy season) (Skujiņš and West, 1973). Durrell and Shields (1960) isolated 41 fungal species from Mojave Desert soils. The dominant genera were *Fusarium, Penicillium, Stemphylium,* and *Phoma,* whereas in Death Valley, the prevalent genera were *Aspergillus, Curvularia,* and *Stemphylium* (Hunt and Durrell, 1966).

Fungal genera present in the cryptogamic crusts are usually *Rhizopus, Mucor,* and certain forms of *Botrytis* (Fletcher and Martin, 1948). The presence of these organisms in the surface soils may indicate the loss of previously present crust by erosion.

The presence of myxomycetes in desert soils was reported by Evenson (1961), who isolated *Badhamia macrocarpa, Perichaena depressa,* and *P. vermicularis* from Arizona desert plant material and considered them common in these soils.

Higher macroscopic fungi, the so-called mushrooms, are rare in the desert areas. However, Long and Miller (1945) described a xeric *Coprinus* in the southwestern United States. Grenot (1974) also demonstrated the presence of

Coprinus arenarius in the Saharan desert, Ougarta mountains. *Fomes robustus,* which is parasitic on seguaro, chollas, and other desert plants, was reported by Davidson and Mielke (1947). Also of interest are the arid-soil endedaphic mushrooms, developing under the soil surface, for example, the truffles, *Tuber* spp., noted by Khudairi (1969) from Iraq. *Terfezia leonis,* and *Tirmania africana,* mycorrhizal on *Helianthemum* spp., have been described from the Saudi Arabian (Trappe, 1981) and the Negev (Rayss, 1959) deserts.

The fungal numbers in desert soil are generally lower than in nondesert regions, but the commonly found genera and species are distributed worldwide. When Mouchacca and Joly (1974, 1976) extensively studied the behavior of *Penicillium* and *Aspergillus* in arid soils, they concluded that the distribution of the various species of *Penicillium* was qualitatively homogeneous. The quantitative variations were a function of certain soil characteristics. The locality and the texture of soils did not affect species distribution. On the other hand, soil reclamation associated with the decrease in soluble salts and enrichment in total nitrogen modified the relative abundance of certain species. The distribution of *Aspergillus* spp. also exhibited only quantitative variations similar to *Penicillium.* The absence of a specific species in a soil apparently had no qualitative significance distinct from its presence in another soil. The taxonomic distribution was insignificantly affected by the soil texture, although regional localization exerted a certain influence.

Given these generalizations, however, the generic and species composition of arid-soil mycoflora is distinct for each desert. The desert mycoflora is dominated by fungi with dark-colored hyphae and other structures. Many pigmented genera and species of Dematiaceae, Sphaeropsidales, and Mycelia sterilia are components of a characteristic desert mycoflora.

Nicot (1960) delineated the general features of two prevalent biological types of fungal adaptation to desert conditions. The first type is characterized by dark, pigmented mycelia and fruiting bodies. The pigments apparently act as solar light filters. Hyphae of many xeric fungi are thick-walled and constitute an additional light-reducing device. A second type of fungi show rapid development on germination, utilizing short periods of favorable growth conditions. They are usually organisms with multicellular spores representing a considerable reproductive capacity. Durrell and Shields (1960) examined the effect of dark pigments as sources of protection against excessive solar radiation. On irradiation at 253.7 nm wavelength, the survival time of *Stemphylium ilicis,* a characteristic desert fungus, was 60 min, as against 2 min for equally dark conidia of *Aspergillus niger.*

An examination of temperature tolerance of fungi (Borut, 1960), conducted on several fungal species isolated from Negev desert soils, showed that most species had a growth optimum at approximately 26°C, increasing for several species to approximately 30°C. The temperature optimum for *Asper-*

gillus fumigatus and *A. niveus* was at 36°C. At 40°C, most species did not grow. These data coincide with the observed phenomenon that there are only relatively few thermophilic microorganisms in desert soils.

Similar to their functions in other ecosystems, the main physiological functions of fungi in desert soils are their ability to decompose organic matter (litter) rapidly and their contribution to the soil structure formation (Borut, 1960; Khudairi, 1969; Went and Stark, 1968).

3.3. Spatial Distribution

More than in any other ecosystem, microorganisms in desert soils exhibit certain stratification in their location, horizontally as well as vertically.

The edaphic and climatic characteristics of arid areas are conducive to the growth of perennial plants in spatially organized groups, the "fertile islands," which have developed over centuries as the result of floral and faunal activities (Garcia-Moya and McKell, 1970; Charley, 1972; Wallace and Romney, 1972; Charley and West, 1975). The organic matter content in fertile islands may reach more than 2%, as compared with less than 0.5% in soil between the concentrations of plant canopies. In response to the organic matter availability, the microbial activities are significantly higher under the plant canopies than in the plant-free soil. Consequently, in examining the microbiology of desert soils, one should consider the two localized environments: one is the canopy-dominated soil; the other is the plant interspace ("bare") soil. Rarely have these distinctions been made in the early literature.

Vollmer *et al.* (1977) examined Mojave desert soils under *Lycium andersonii* canopies and in the interspace. Bacterial numbers reached 8×10^6/g soil under the canopy and decreased in the interspace, where the highest numbers were 3.8×10^6/g. Similarly, actinomycetes reached 9.2×10^6/g under the canopy and decreased in the interspace. Nicot (1960) reported that fungal microorganisms concentrated at certain depths below the plant canopies, and Dubost and Hethener (1966) found that the highest numbers of soil microorganisms existed at a depth of 30 cm in the soils of Tassili N'Ajjer. The stratification was especially pronounced during dry seasons, as demonstrated by Elwan and Diab (1970d), who found that bacteria were more numerous in the deeper Arabian desert soil during the dry than during the rainy season. Often in desert soils, hyphal concentrations may exist at a depth between 10 and 30 cm, as observed also in the Sonoran and Mojave deserts. Cameron and Blank (1965) reported mean fungal propagule numbers an order of magnitude higher (34,000/g) at the 30-cm depth of six California desert profiles than in the surface (3700 propagules/g) or other soil horizons.

The most important spatial distribution factor of microorganisms, however, is the presence of roots, there being a significant increase in microbial numbers in the rhizospheres (cf. Section 5.2). Distinct from other environments, the microbial activity in arid soils is usually concentrated on the soil surface. The surface is often covered by cyanobacteria and lichens forming the cryptogamic crusts, among other characteristic biological surface features (cf. Section 4.1).

3.4. Temporal Variability

The rhythm of seasonal climatic conditions causes considerable variation in microbial numbers and activities. Killian and Feher (1935, 1938) demonstrated that bacterial numbers increased from 16,000 to 1.5×10^6/g soil during the rainy period in a desert soil. Similarly, in an oasis soil on increase of soil water content, the fungal numbers increased from 30,000 to 650,000 propagules/g. In a particular Saharan soil, as the soil water content increased from 1.5 to 10%, the microbial numbers increased from 0.1×10^6/g to about 1.8×10^6/g. Elwan and Diab (1970d) found a similar variation in the rhizosphere soils of several plant species.

In her report on microbial presence in several Chilean soils, Franz (1971) demonstrated considerable seasonal and annual variation of microbial numbers in soils. In the succulent-dominated semidesert soils near Quebrada de la Plata, the aerobic bacteria numbers varied between 300,000 and 5×10^6, depending on the season. The variation between consecutive years was also about 3-fold. Similarly, actinomycetes varied from 20,000/g in the dry season to 200,000/g in the wet season, representing a 10-fold increase. Similar changes were also evident with fungi and with anaerobic bacteria. The changes occurred not only in the surface 5-cm soils, but also in the 25–30 cm horizon. The seasonal increase of total aerobic bacteria during a rainy season in the northern Egyptian desert littoral calcareous dunes was over 10-fold, from 50,000 to more than 500,000 bacteria/g, whereas in the inland desert, the increase was 3-fold in the surface 25-cm soils. Similar increase ranges were present for actinomycetes, spore-formers, and also in subsurface (25–50 cm depth) soils. The values for Egyptian soils were based on 3-year averages (Abdel-Ghaffar and Fawaz, 1977).

Although the literature usually notes the increase of microbial numbers during periods of water availability, these values also represent the death rates of arid-soil microorganisms on soil desiccation during dry periods. The very high death rate of organisms in the dry calcareous dune soil is particularly striking. It exceeds comparable microbial death rates in temperate agricultural soils on air-dry storage for a comparable 10-month period.

3.5. Effect of Soil Water Content

Microbial numbers and activities diminish greatly when the water content in soil decreases. A considerable amount of information has been collected on microbial growth when the water potential, ψ_w, becomes limiting (Table II).

Table II. Approximate Limiting Water Potentials for Microbial Growth at 25°C[a]

Water potential (MPa)	Water activity (a_w)	Reference points	Bacteria	Yeasts	Fungi
−0.003	1.00	Blood	*Caulobacter*	—	—
−1.5	0.989	Wilting point	*Spirillum*	—	*Sclerotium rolfsii* (sclerotia)
−3.0	0.978	Seawater	Chemoautotrophs	—	—
−7.1	0.95	—	Most gram-negative rods	Basidiomycetous yeasts	Basidiomycetes
					Rhizopus nigricans
−15	0.90	—	Most cocci	Ascomycetous yeasts	*Fusarium*
			Lactobacillus		Mucorales
			Bacillus		*Aspergillus flavus*
−22	0.85	—	*Staphylococcus*	*Saccharomyces rouxii* (in salt) *Debaryomyces*	*Aspergillus niger* *Paecilomyces varioti*
−31	0.80	—	—	—	*Penicillium*
−39	0.75	Salt lake	Halobacteria Halococci	—	*Sporendonema sebi* *Aspergillus* spp. *Wallemia*
−49	0.70	—	—	—	*Chrysosporium* *Aspergillus ruber*
−59	0.65	—	—	—	*Erotium*
−70	0.60[b]	—	—	*S. rouxii* (in sugars)[b]	*Xeromyces bisporus*
−82	0.55	DNA disordered	—	—	—

[a]Adapted from Brown (1976), Griffin (1972), and Kushner (1978). [b]Value by Brown (1976); Kushner (1978) assigns $a_w \approx 0.85$ for *S. rouxii* in all media.

Generally, fungi proliferate and initiate growth at lower water potentials than do bacteria. The lower ψ_w limits reflect the fungal ability to proliferate at water potentials when most bacterial activities in arid soils would be limited, indicating that fungi may be contributing more to decomposition in "dry" soils than do bacteria. It has been suggested that at water potentials greater than −0.1 megapascal (MPa), bacteria grow actively in the pores of well-aggregated soils and fungi would be generally excluded. At $\psi_w > -0.1$ MPa, however, various fungi would be able to enter the aggregates, whereas bacterial activity at these water potentials would be limited (Cook and Papendick, 1970).

Dommergues (1962b) noted that in horizons between 5 and 20 cm depth in western North Africa, the soils remained between pF 4.2 and 5 (−1.5 to −10 MPa) for an extended period of time following a rainy period. At these soil water potentials, the plants approached the senescent stage (permanent wilting point corresponds to pF 4.2 or −1.5 MPa), whereas microbial activities were evident up to pF 5.6 (−40 MPa). Dommergues classified desert-soil microorganisms according to their response to water availability in three groups. The threshold value for *hyperxerophilic* microorganisms is greater than pF 4.9 (-8 MPa). According to Dommergues (1962b), these organisms included bacteria participating in tyrosine and casein ammonification, glycolysis, glycerophosphate mineralization, amylolysis, and decomposition of plant residues. The threshold values for the second group, the *xerophilic* microorganisms, lie between pF 4.9 and 4.2 (-8 and -1.5 MPa). These organisms participated in cellulose decomposition, casein ammonification, nitrification, and partially in amylolysis. The third group are the *hydrophilic* microorganisms, having a threshold value below pF 4.2 (-1.5 MPa). These are carboxymethylcellulose decomposers and nitrogen fixers, for example, *Azotobacter chroococcum*. Their water demands equaled those of plants.

When soil loses water progressively, incrementally selective changes of microbial activities occur (Dommergues, 1962b). During the drying of soils, microbial activities show certain stages of a decreasing but variable intensity of mineralization, resulting in an accumulation of organic matter and nitrogen. Depending on the composition of the microbial community, the soil type, and the nature of organic matter, mineralization of carbon compounds, such as cellulose and lignin, begins at water potentials considerably above the permanent wilting point, whereas the respective bioses can be decomposed at water potentials below -1.5 MPa. Nitrogen mineralization may also take place at lower ψ_w values. Thus, as the soil dries out, the activity of microorganisms decreases and a qualitative modification of processes of organic-matter decomposition occurs, as certain groups of organisms are more rapidly inactivated than others. Near the permanent wilting point, for example, cellulose decomposition would be more prominent than mineralization of nitrogen, and on further drying, ammonification would prevail over nitrification.

Many microorganisms in desert soils are not spore-formers, nor do they otherwise form any resting structures. Elwan and Diab (1970d) showed that in very dry sands, which contained 0.95–2.5% water and which were also poor in organic matter (0.09–0.033% organic C), at least 84% of the bacteria were in a vegetative form. They suggested that moisture from dew was possibly sufficient to maintain the active state of bacterial cells for a few hours after sunrise. Rychert and Skujiņš (1974a) reported a similar effect of dew on nitrogen fixation by cyanobacterial crusts, indicating that the dew formation might be an important factor of microbial activity in desert surface soils (Biederbeck *et al.*, 1977). Brock (1975) examined the effect of ψ_w on the growth and photosynthesis of *Microcoleus,* a desert crust component, and demonstrated that the growth was partially inhibited at -0.7 MPa and completely inhibited at -1.8 MPa, whereas the photosynthesis was completely inhibited at -2.8 MPa. The organism was more sensitive to the reduction of ψ_m than to that of ψ_s. Brock concluded that *Microcoleus* was poorly adapted for growth at low water potentials, but had a considerable ability to survive severe drought conditions.

3.6. Effect of Temperature

Although many organisms in desert soils may survive surface soil temperatures reaching higher than 70°C, there are few thermophilic organisms present. Pochon *et al.* (1957) demonstrated that in Saharan soils, there were 100–600 thermophilic bacteria/g, and no nitrogen-fixing and nitrifying thermophiles were present. Ammonification, denitrification, cellulolysis, hemicellulolysis, and amylolysis were negligible at thermophilic temperatures. As the temperature increased, the relative abundance of bacteria with simple nutrient needs also increased, whereas bacteria requiring more complex nutrient media were favored by lower temperatures (Elwan and Diab, 1970b; Diab, 1978).

The ecological significance of finding few thermophilic organisms in desert soils reflects the environmental conditions in deserts, where increased insolation and temperature at the same time also desiccates the soil. Consequently, microbial growth in desert soils may occur mostly at lower temperatures coinciding with prolonged availability of water. On increase of the temperature, soil dries out rapidly, reaching $\psi_w > 40$ MPa, and no significant microbial proliferation may be expected.

3.7. Adaptation to Arid Environments

Specific microbial populations have been well researched in certain harsh environments, e.g., halophiles in saline lakes, thermophiles in hot springs, barophiles on the oceanic floors, and others. The microbial populations reflect adaptation and natural selection to these environments. It is not yet clear whether a specific microbial community exists, adapted and selected to environments characterized by prolonged periods of low water potentials in desert soils. However, it is reasonable to generalize that species diversity declines as environmental adversity increases. Bacterial species diversity in desert soils has not been significantly studied, but the low microbial diversity in the dry valleys of Antarctica (Cameron, 1972) is notable.

In general, processes and mechanisms for microbial population selectivity and adaptation to arid environments have been poorly studied (Alexander, 1976). Most such studies have been directed to thermophiles and halophiles (Heinrich, 1976; Shilo, 1979). There are indications, however, that certain mechanisms might be present to effect selectivity and adaptation of microflora to arid soil environments, in which water is the primary limiting factor. During incubation of soils at discrete water activities, soil ATP reached the highest concentrations at -1.8 MPa in arid soils, whereas in mesic forest soils, the highest concentrations were present at less than -0.2 MPa (Knight and Skujiņš, 1981). Observations that arid-soil microorganisms often exhibit the highest physiological activities between -0.5 and -0.7 MPa may also indicate certain selectivity. Similarly, nitrification was observed to have a temperature maximum at 35 °C in Arizona and New Mexico soils, but at 20–25 °C in northwestern United States soils (Mahendrappa *et al.,* 1966). Focht and Martin (1979), however, state that the microbial community of semiarid agricultural soils is basically similar to that of more humid soils. One should note that microorganisms in temperate and humid soils are subjected to drastically lowered water potentials in the soil microenvironments during intermittent soil drying. Presumably the organisms have developed certain defense mechanisms for survival. Examination of such aspects has not been sufficiently addressed in soil microbiology, although the empirical environmental factors influencing microbial death rates in temperate and humid soils on drying are well known (Dommergues and Mangenot, 1970).

The 1978 Dahlem Conference (Shilo, 1979) reviewed the current knowledge about physiological and biochemical adaptations of microorganisms to extreme environments. Many of the aspects discussed were pertinent to our understanding of natural selection, adaptation, and function of microorganisms in desert soils (Horowitz, 1979; Griffin and Luard, 1979; Brown, 1979; Lanyi, 1979).

Although a considerable amount of descriptive information has been collected regarding microbial activities in arid environments, little is known about their physiological and biochemical functions, particularly at water potentials below −5 MPa. Griffin and Luard (1979) noted that experiments on "interrelationships between growth, ionic transfer, osmotic regulation, and respiration are likely to be particularly rewarding." In addition, information on microbial morphological and physiological defenses against desiccation and on energy expenditure in water assimilation and retention under low water potentials is also of interest.

The question has been raised whether an abiotic, sterile soil exists any place in the world. Rougieux [(1966) (cited in Binet, 1981)] found a soil in a reg devoid of vegetation and containing only 90 bacteria/g, without actinomycetes or fungi. In conjunction with the proposed extraterrestrial life exploration, considerable attention has been devoted to examining the presence of life in extreme environments on earth, specifically in Antarctica. Although there are spurious reports that sterile soils have been found [e.g., Asgard Range, Antarctica (Cameron, 1969)], they usually pertain to subsurface horizons [for example, subsurface sample 542, reported by Cameron (1972)]. Various organisms, especially photoautotrophic algae and cyanobacteria and a few bacterial species, have been found proliferating either endolithically or under the ice cover in the physiologically extreme Antarctic environments. Evidently, the first invaders of any extreme environment are certain autotrophic organisms, including nitrogen-fixers, which would be accompanied by heterotrophic bacteria and other organisms. Such studies have indicated the great adaptability of microorganisms to survive and to proliferate in extreme environments (Llano, 1972; Heinrich, 1976). It is also expected that aeolian distribution of microorganisms would universally "contaminate" any environment.

4. Biology of Desert Soil Surface Features

A considerable portion of global deserts are covered by hardened surfaces, variously named "desert pavements," "desert crusts," or "desert varnishes." *Desert pavements* are characterized by accumulation of coarse sand, gravel, and stones on the soil surface as a result of wind and water erosion following removal of sand and clay particles. These materials are often cemented or encrusted with various salts, calcium carbonate, gypsum, and silicate. If the hardened particles are coated with iron and manganese oxides, giving them a polished appearance, the term *desert varnish* is used. Hardened desert soil surfaces may also be formed following occasional cultivation of soils in years with increased rainfall. The soil is compacted at plowing depth, and during a subsequent dry season, the loosened, sandy layer is removed by aeolian erosion, exposing the hard, *compacted* soil layer. The hardened desert soil surfaces are often inhabited by cyanobacteria, algae, and lichens, forming *cryptogamic crusts*. The crusts are usually complex microbial communities, and various fungi and bacteria are present.

4.1. Cryptogamic Crusts

Hardened porous dense crusts underlain by a flaky microhorizon are often formed on sandy and silty surfaces. Such crusts are characteristic of arid soils in temperate climates, although they are also present in tropical and subtropical deserts (J. T. Miller and Brown, 1938). The physical and chemical processes contributing to crust formation are not clear. The presence of calcium carbonate and seasonal fluctuations in hydrothermal regime are important factors in crust formation, and the process is possibly enhanced by the presence of plant roots (Kovda *et al.,* 1979). The major contributing factor to cryptogamic crust formation is the presence of fungi, lichens, algae, and cyanobacteria. On sandy soils, the cryptogamic crusts are fragile. More stable crusts with increasing presence of desiccation-resistant lichen flora develop on clay and silt-containing soils. Cryptogamic crusts are usually absent in sandy desert soils subjected to aeolian sand movement. Many cyanobacteria are nitrogen-fixing, and cyanobacteria and algae act as primary producers and provide carbon input in the surface soils. As a result, cryptogamic crusts are associated with a wide range of various bacteria and fungi. The fungal hyphae form a

certain continuum in the soil to a depth of several centimeters and contribute
to crust stabilization. Cyanobacteria present in the crusts usually include *Oscillatoria, Nodularia,* and *Microcoleus.* Common components of the cryptogamic
crusts include various green algae and lichens (Friedmann and Galun, 1974).

Three examined types of crusts (cryptogamic, pavement, and compacted)
in the Presaharan Desert in Tunisia (Skujiņš, 1977b) had a similar type of
microbial community. The total number of microorganisms in the crusts
increased as compared to the microbial numbers in surface soils in grazed areas
without evidence of surface crusts (Table III), but the species diversity in
crusts decreased. The dominant fungal genera in crusts were *Alternaria,
Fusarium,* and *Phialomyces,* whereas in noncrusted soil surfaces, the dominating genera were *Alternaria* and *Penicillium,* followed by *Fusarium.* The aerobic bacterial diversity in crusts decreased significantly, dominated by coryneforms, constituting 90% of the population, whereas in noncrusted surface soils,
coryneform population fluctuated between 30 and 60%. Cyanobacteria *(Oscillatoria)* were present but not dominant in these crusts. On the surfaces of stabilized Tunisian soils, Englund (1975) found species of *Calothrix, Tolypothrix,
Anabaena,* and *Nodularia,* with a number of *Nostoc* species as dominating
cyanobacteria. In dry sandy soils, the fixation rates reached 75 ng N_2/g per hr,
but no N_2 fixation or cyanobacteria were found in sand dunes. In the cryptogamic crusts of the Great Basin Desert, *Nostoc, Lyngbya, Microcoleus, Phormidium,* and *Oscillatoria* were prevalent and dominating (Klubek and Skujiņš,
1980). Renaut and Sasson (1970) list 40 species representing *Chroococcus,
Merismopedia, Calothrix, Tolypothrix, Nostoc, Anabaena, Cylindrospermum,
Spirulina, Oscillatoria, Phormidium,* and *Lyngbya* from arid Moroccan soils.

4.1.1. Significance of Cryptogamic Crusts

Fletcher and Martin (1948) demonstrated that certain cryptogamic
crusts, reaching a thickness of 2.5 mm or more, increased the tensile strength
of soil, reduced erosion, and increased soil organic matter content.

Further studies on this subject suggest that cryptogamic crusts increase
soil stability and water infiltration and reduce erosion and runoff. Many
reports, however, merely note the influences by implication, and a hard data
basis on the effects of cryptogamic crusts on various soil characteristics is
scarce. Gifford (1972) compared sites with and without cryptogamic cover and
found that sites with cover had significantly higher infiltration rates. Sorensen
and Porcella (1974) noted that erosive loss of nitrogen took place where cryptogamic crusts were removed. Booth (1941) and Fletcher (1960) found that
the removal of crusts considerably increased runoff. Dogan (1975) reported
that sedimentation doubled in areas where the crust was removed as compared
with areas of intact crust. Dregne (1968), in an appraisal of research on surface
covers of desert soils, stated: "Whether algae and lichen are significant factors

Table III. Microbial Numbers (per g × 10³) in Several Types of Crust, Presaharan Desert, Tunisia[a]

Crust type	Aerobic bacteria	Proteolytic bacteria	Microaerophilic bacteria	Actinomycetes (propagules)	Fungi (propagules)
Grazed area (0–3 cm)[b]	810	150	0.8	220	2.8
Crust					
Silty surface (0–1 cm)	3,040	54	7.6	430	7.0
Clay sediment (0–1 cm)	41,000	3,000	0.9	30,000	110
Exposed compacted surface formed after plowing (0–1 cm)	310	100	1.5	430	1
Calcareous pavement	540	ND	ND	ND	8.3

[a]Derived from Skujiņš (1977b). (ND) Not determined.
[b]Crust formation disrupted by animal trampling.

in soil stabilization in upland soils is unresolved as is effect of such crusts on water penetration and runoff, to say nothing of the effect on soil development." Faust (1971), for example, found no detectable differences in runoff, infiltration, or sediment yields from soils covered with cyanobacteria and soils denuded of cyanobacterial growth. Although information on the effects of cryptogamic crusts on soil stability is sparse and in some cases conflicting, the majority of the available data do suggest that these crusts play an important role in increasing soil stability and related soil parameters.

Although the N_2-fixing ability of cyanobacteria, either free-living or associated with fungi (lichen), has been documented and extensively discussed, their importance to the nitrogen budget in desert ecosystems is not well defined (Skujiņš, 1981). The amount of N_2 fixed by cyanobacteria and lichen crusts in desert ecosystems is a function of population characteristics of N_2-fixing organisms, water availability, and climatic conditions. Hence, the amount of N_2 fixed can be variable, both temporally and spatially. Some representative N_2-fixation rates for desert areas include 1–18 kg/ha per year in the Sonoran Desert, 7–16 kg/ha per year in the Great Basin Desert, and 0.5–1 kg/ha per year in the northern Mojave Desert (Rychert et al., 1978; Skujiņš, 1981). In localized, well-developed crust situations under favorable climatic conditions, the rates may reach 30 mg N/m^2 per day (Rychert and Skujiņš, 1974a). Following precipitation, the desiccated cyanobacteria cells recover active physiological functions within 15–60 min. They may photosynthesize at water potentials of −1 megapascal (MPa) or less, but apparently they fix N_2 only when the water potential in the ambient microsite is not less than −0.2 MPa. There is also evidence indicating that during a dry season, they may respond to dew (Rychert et al., 1978), with the highest nitrogenase activity during morning hours, before drying. N_2 fixation reaches optimal rates at light intensities equivalent to daylight with a heavy cloud cover; consequently, their activity is not limited by decreased light intensities during rainy and cloudy periods. Following a rainfall, 3–4 g N/ha per hr may be fixed in the Sonoran Desert (MacGregor and Johnson, 1971).

The variation in N_2-fixation rates probably reflects a typical arid ecosystem heterogeneity, and the average annual rates depend on the localized variability of cryptogamic ground cover. Cryptogams cover 40–80% of the ground in the Curlew Valley of the Great Basin Desert, whereas the coverage decreases to 0.1% in the northern Mojave Desert (Lynn and Cameron, 1973; Wallace et al., 1978). Next to soil characteristics, the variation of cyanobacterial cover is influenced by climatic conditions and by allelopathic effects of prevailing vegetation (Rychert and Skujiņš, 1974a,b).

N_2 fixed by cyanobacteria may be assimilated by angiosperms (Mayland and McIntosh, 1966; Stewart, 1967; Snyder and Wullstein, 1973). In the Great Basin Desert, however, most of the N_2 fixed by cyanobacteria is rapidly lost by denitrification, and little nitrogen is added to soil (Skujiņš and Klubek, 1978).

The cryptogamic crust-fixed N_2 accumulates in the few surface centimeters of soil. Much of it becomes lost from the soil again during the same favorable climatic conditions conducive to increased microbial activities. As the photoautotrophic cyanobacteria in the crust fix N_2 and CO_2 and increase in biomass, a part of the crust organisms die and decompose. The fixed N_2 is released as NH_4^+ in the soil, and most of it is nitrified to NO_2^- and NO_3^-. These components may be reduced to N_2 or N_2O, and the fixed nitrogen is lost from the soil. Organic matter, resulting from the primary production by cyanobacteria, serves as an energy source for denitrification. Nitrification and denitrification may take place in the same crust microenvironment (Skujiņš, 1981). Conversely, in the northern Mojave Desert, where the input of nitrogen in soils by cyanobacterial fixation is low, the rate of nitrogen loss is also slow as compared to the Great Basin Desert, and the soil nitrogen pool adequately meets the needs of seasonal plant productivity (Wallace et al., 1978). Wallace and co-workers concluded that although N_2-fixation rates in the Mojave Desert by cryptogams are low, fixation may contribute significantly to the nitrogen cycle.

4.2. Desert Varnish

Biological formation of iron- and manganese-oxide-containing desert varnish has been suggested by Bauman (1976) among other authors. Dorn and Oberlander (1981) have verified the presence and participation of *Metallogenium*- and *Pedomicrobium*-like bacteria in the formation of desert varnishes. The organisms concentrate ambient manganese that becomes greatly enhanced in brown and black varnish. Nevertheless, Elvidge and Moore (1979) have proposed a purely chemical model of manganese varnish formation in deserts.

According to Dorn and Oberlander (1981), the microbial origin of desert varnish does not exclude the possibility that natural varnish films can be formed without biological assistance. However, a purely physicochemical origin seems unlikely, and the field observations support the microbial origin.

4.3. Lithic Organisms

Many organisms in desert environments exist and proliferate on the surface or within exposed rocks and stones. The most common in these microenvironments are photoautotrophic cyanobacteria and algae and their associations with fungi, the lichens. Since many arid-environment cyanobacteria are nitrogen-fixing, they are self-sufficient and independent from external energy and carbon sources.

4.3.1. Fungi

Microcolonial structures of fungi on desert rocks exposed above the soil and in the absence of detectable lichen or algal growth have been reported by Staley et al. (1982). On examination of rocks from desert pavements from the Australian, the continental Asian, and the Sonoran and Mojave deserts, the prevalent fungi were identified as belonging to Capnodiales, and most others were Hyphomycetes (Fungi Imperfecti) such as *Taeniolella subsessilis* (Ell. and Everh.) Hughes, and *Bahusakala*- and *Humicola*-like organisms. Since the rock surfaces showed high respiration rates but no photosynthesis, the authors concluded that there was no carbon dioxide fixation by primary production. The authors suggested that these organisms must rely for growth on an external source of organic nutrients, such as windblown material brought to the rock surface from the surrounding soil and vegetation. During the wet period, this material might be used as nutrients for the growth of the fungi.

4.3.2. Algae and Cyanobacteria

The variety of desert habitats in which algae and cyanobacteria occur has been excellently described by Friedmann and Galun (1974). Friedmann et al. (1967) and Friedmann and Galun (1974) classified the desert algae and cyanobacteria habitats and distinguished between the edaphic (soil) and the lithophytic (rock) algae. The authors list over 50 genera of algae found within desert soils (endedaphic algae), including representatives of Xanthophyceae and diatoms; in addition, they reported 20 genera of endedaphic cyanobacteria. It has been speculated that the abundance of algae in soils without higher plant vegetation may be due to the higher water-retention facility of soils in the absence of vegetation. It is evident, however, that plants may exert strong allelopathic influences on the prevalence of cyanobacteria (Rychert and Skujiņš, 1974b; Rychert *et al.,* 1978).

Allelopathic influences may also regulate the distribution of algae and cyanobacteria living on the surface of desert soils (the epedaphic algae, either free-living or forming lichen associations, and described as cryptogamic crusts). Although the literature on epedaphic algae is extensive, the intrinsic nature of lichenization has been poorly investigated (Friedmann and Galun, 1974). The epedaphic community is dominated by cyanobacteria. The common genera are *Schizothrix, Microcoleus, Oscillatoria, Anabaena, Gloeocapsa, Phormidium, Lyngbya, Nostoc, Calothrix, Plectonema, Nodularia,* and *Aphanothece,* among others. Numerous epedaphic genera of green algae have also been described as components of cryptogamic crusts, such as *Chlamydomonas, Chlorococcum, Chlorella, Cystococcus, Scenedesmus,* and others.

The hypolithic algae grow on the lower surfaces of translucent stones, partially buried in soil. These organisms also use a variety of calcareous objects of animal origin as habitats. The hypolithic algae have been described for most deserts, and an excellent ecological study of the hypolithic habitat has been done by Vogel (1955). The hypolithic flora is highly diversified and comprises filamentous as well as coccoid cyanobacteria and green algae (Friedmann and Galun, 1974).

Two types of endolithic (rock) algae may be distinguished with respect to their habitat. The *chasmoendolithic* organisms inhabit rock fissures and cracks, and the *cryptoendoliths* inhabit internal structural cavities of porous rocks. The endolithic microbial communities depend on photosynthetic primary production, and only translucent rocks are suitable as growth habitats. The dominant endolithic organisms are cyanobacteria, such as *Gloeocapsa,* isolated from various rocks from the Negev Desert, Death Valley, and Sonoran Desert, and from Antarctica (Friedmann and Ocampo, 1976), and apparently *Chroococcidiopsis,* isolated from sandstones in an Antarctica dry valley (Friedmann, 1982). The cyanobacteria in rocks are accompanied by bacteria. The lichens are also present as cryptoendolithic organisms in Antarctica (Friedmann, 1982). These lichens may often be exfoliated, and filamentous fungi and unicellular green algae, most commonly *Trebouxia,* may be present. The endolithic cyanobacterial flora is similar in hot and cold desert soils (Friedmann and Ocampo, 1976).

The endolithic organisms participate significantly in rock-weathering processes. The environmental conditions for endolithic microbial growth are regulated by the availability of water and nitrogen, among other nutrients. Although most of the nutrients may be extracted from the rocks by the various organisms, water remains a significant limiting factor. Lichens, however, may utilize atmospheric water vapor and dew (Lange *et al.,* 1970a,b), and it is likely that cyanobacteria can do the same (Rychert and Skujiņš, 1974a). Although many cyanobacteria fix atmospheric nitrogen, Friedmann and Galun (1974) suggest that *Azotobacter*-like organisms may be of equal importance in lithic environments.

5. Microbial–Plant Associations

5.1. Mycorrhizae

The associations of mycorrhizae (Phycomycetes, Endogonaceae family) with higher plants have been reviewed by a number of investigators (Marks and Kozlowski, 1973; Sanders *et al.*, 1975; Gerdemann and Trappe, 1974). Trappe (1981) listed 264 plant species from arid and semiarid environments that have been examined for mycorrhizal presence on roots, and approximately 25% of the species have exhibited associations with fungi. Trappe (1981) cautioned that the information for many species was based on a single observation and that certain species reported as mycorrhizal were noted as nonmycorrhizal by other authors. Woody species, forming ectomycorrhizal associations, mostly from Fagaceae, Pinaceae, Myrtaceae, Rosaceae, and Salicaceae families, are usually absent in desert environments. Most species of Asteraceae, Poaceae, Fabaceae, Rosaceae, and Solanaceae usually form vesicular–arbuscular (VA) endomycorrhizal associations in arid habitats. The Apiaceae and Zygophyllaceae species are usually nonmycorrhizal. Although some Chenopodiaceae and Cactaceae species exhibit mycorrhizal associations, most appear to be nonmycorrhizal. Annuals of typical mycorrhizal-associated plant families are probably as receptive to VA mycorrhizal infection as perennials (Poma, 1955). Incidence of mycorrhizal colonization in an arid environment may vary with season, with soil moisture availability (Staffeldt and Vogt, 1975), and with the composition of the plant community (Hirrell *et al.*, 1978). VA endomycorrhizal associations generally exhibit low fungal–plant species specificity. Viable VA fungi persist in roots of perennial hosts over the dormant season, but annuals depend on the presence of propagules in the soil for new colonization each growing season. In plant communities of annuals mixed with perennials, the hyphae present in perennial roots may serve as inocula to colonize annuals. The climatic conditions are undoubtedly the most significant factor in initiating mycorrhizal infections. Trappe (1981) noted that VA endomycorrhizae indigenous to arid soils probably have evolved certain defenses against natural hazards. Unfortunately, little is known about the species distribution, specificity, and ecology of VA mycorrhizal fungi in arid environments (Mosse *et al.*, 1981). Although the specificity of plant–fungal species interaction is low, the fungi differ in their ecological requirements and adaptability to the higher plants. Transfer of spores from one habitat to another may not ensure mycorrhizal infection of the same plant species in the new habitat. Some VA mycorrhizae indigenous to desert soils are unique to that environment, for example, *Glomus deserticola* (Trappe, 1981).

One may assume that fungal hyphae exert a phycosphere effect on microorganisms similar to the rhizosphere effect exerted by plant roots. Research on these relationships, especially in arid soils, is practically nonexistent. An intriguing question is whether microbial associations with fungal hyphae may induce an associative type of nitrogen fixation. Rougieux (1963), however, noted a substantial increase in microflora near sporocarps of *Terfezia boudieri*. *Azotobacter chroococcum* was present only near sporocarps, and the extracts of the fungus stimulated growth of the *Azotobacter* cultures. *Azotobacter* may also envelop spores of the mycorrhizal fungus *Glomus fasciculatus* (Gerdemann and Trappe, 1974) and may be stimulated by the fungal presence in soils (Bagyaraj and Menge, 1978). It is well known that mycorrhizal presence enhances nodule formation and nitrogen fixation of Fabaceae species, particularly in phosphate-deficient soils (Mosse *et al.*, 1981). Again, such interactions in arid desert soils have not been examined, although inoculation of arid-soil rangeland plants by mycorrhizae has recently drawn considerable attention, especially in areas requiring revegetation. It has been demonstrated that the absence of mycorrhizae greatly reduced the competitive ability of normally mycorrhizal plant species, especially in competition for phosphorus (Hall, 1978). Under high soil temperature, present in arid soils, mycorrhizae appeared critically important for phosphate nutrition of mycorrhizae-dependent plants (Barrow *et al.*, 1977).

During revegetation of severely disturbed soils, nonmycorrhizal species strongly dominated the primary succession. Less than 1% of colonizing plants on the disturbed site were mycorrhizal, although 99% on the adjacent control sites were mycorrhizal (Reeves *et al.*, 1979). These authors speculated that mycorrhizal species were more effective competitors for soil water and for nutrients than were the nonmycorrhizal species. During the slow succession process characteristic of arid habitats, the mycorrhizal species gradually replaced the nonmycorrhizal plant species as the fungi colonized the soil. R. M. Miller (1979) demonstrated that in disturbed plant communities in the Red Desert of Wyoming, no plants were infected with mycorrhizae, whereas plants present in the neighboring undisturbed community showed a 90% infection rate. Inoculation of a disturbed soil with mycorrhizal fungi apparently promotes revegetation. In addition to mycorrhizal involvement in phosphate nutrition of plants in semiarid regions, the VA mycorrhizae may assume particular significance for enhancing water availability to plants (Nicolson and Johnston, 1979; S. E. Williams and Aldon, 1976), but their presence also increases evapotranspiration rates (Allen *et al.*, 1981). However, there is still insufficient knowledge to adequately understand the role of mycorrhizae in water relationships of higher plants. The importance of mycorrhizal presence for plant growth in nutrient-poor arid soils, particularly in land disturbed by strip min-ing, has been emphasized by Aldon (1978), Trappe (1981), Hall and Armstrong (1979), Lindsey et al. (1977), and Cundell (1977).

5.2. Rhizosphere

Plant rhizospheres are important channels of microbial activities in desert soils. Since the rhizospheres provide ample carbon sources for microbial activities in the otherwise organic-matter-poor arid soils, the rhizosphere/soil (R:S) ratios in desert soils are comparatively severalfold higher than similar ratios in more mesic and humid soils, and the rhizosphere effect is more pronounced both quantitatively and qualitatively than in other soils. The rhizosphere effect in deserts was first described in some detail by Vargues (1953) under *Anabasis aretioides* communities in Algeria, and Thornton (1953) attributed the presence of microorganisms in Sonoran desert soils largely to the presence of roots. A pronounced rhizosphere effect in Egyptian desert soils was demonstrated by Montasir et al. (1958a,b, 1959a–c). According to Elwan and Mahmoud (1960) and Mahmoud et al. (1964), *Maltakaea callosa* rhizosphere samples contained 51.1×10^6 bacteria/g, whereas the nonrhizosphere soil contained only 73,000 bacteria/g, giving an R:S ratio of 700. The rhizosphere population contained 600,000 cellulolytic bacteria/g, 150,000 nitrifiers/g, 350,000 *Azotobacter*/g, and 52,000 *Clostridium*/g rhizosphere soil. The spore-formers *(Bacillus subtilis),* however, made up only 0.12% of the total microbial community, and the R:S ratio was 2.6. *Azotobacter* and *Clostridium* apparently were present only in the rhizosphere, not in the root-free soil.

Increased populations of *Bacillus subtilis* and *Aspergillus terreus* in the rhizosphere of *Farsetia aegyptiaca* were ascribed to the production of growth factors by the roots. Conversely, the fungi and bacteria stimulated germination of seeds, probably due to production of various growth factors (Montasir et al., 1958b, 1959a,b). Mahmoud et al. (1964) indicated that the *M. callosa* rhizosphere generally included *Alternaria, Aspergillus,* and *Fusarium,* whereas *Penicillium* was dominant in the plant interspace soil. Seasonal effects on bacterial activity in the *Rhazya stricta* rhizosphere were examined by Elwan and Diab (1970a–c). As expected, seasonal variation had a strong effect on the rhizosphere effect in arid soils. Microorganisms with complex nutrient requirements were present throughout the soil in the winter and spring, but only in the rhizosphere during the summer (Elwan and Diab, 1970b,c). In the rhizosphere of *Artemisia monosperma,* microbial activity decreased in the summer, which coincided with the slowdown of plant metabolic activity (Elwan and Diab, 1970a). In the rhizosphere of *Panicum turgidum,* however, the highest microbial numbers were found in the summer (Elwan and Diab, 1970c). As the temperature increased, the number of bacteria with simple nutritional requirements also increased, whereas those with complex requirements decreased. In the *P. turgidum* rhizosphere, the highest numbers of cellulolytic bacteria were present during the spring, coinciding with the highest rainfall. The authors delineated three types of environmental effects on temporal and spatial developments of microorganisms in arid soils: (1) soil effect, whereby the number of bacterial cells decreased with depth in the spring, but increased in the summer, suggesting an adaptation to the environment; (2) sporulation pattern, representing a general decrease in sporulation with depth in the spring and summer; and (3) rhizosphere pattern, in which the effect of the presence of roots increased with depth in the spring and decreased in the summer (Elwan and Diab, 1970d). They also suggested that bacteria containing mucilaginous coats may in certain cases protect the roots against excessive desiccation.

5.3. Rhizosheaths

A unique structure of xeric plant root systems is the rhizosheaths, cemented agglomerations of sand particles around plant roots, described first by Price (1911) and later by Wullstein et al. (1979), Wullstein (1980), and Wullstein and Pratt (1981). The exterior of rhizosheaths consists of aggregated fine and very fine sand grains. The cemented layer is formed by plant root exudates, and the layer is conducive to concentration of organic matter and to the retention of water. Consequently, the environment within the rhizosheaths increases microbial activity. Sheath thickness rarely exceeds 1 mm and in most cases varies between 0.2 and 0.5 mm. Within the rhizosheath, the plants develop an exceptionally dense mass of root hairs, and root hairs protrude just beyond the surface layer of grains. Rhizosheath interiors are usually populated by various types of bacteria, including *Bacillus polymyxa,* and a fungus, *Olpidium,* but otherwise fungi are mostly absent. Unique are *Ancalomicrobium-* and *Hyphomicrobium*-like organisms present in rhizosheaths as described by Wullstein and Pratt (1981). Present in the rhizosheath environment are nitrogen-fixing bacteria contributing to the nitrogen influx in the desert ecosystem. The rhizosheaths have been described for a variety of plant species, but they are especially pronounced in various Poaceae species.

6. Physiological Activities

6.1. Decomposition

Although desert soils contain most physiological and taxonomic groups of microorganisms, the range of activities of these microorganisms is limited by adequate moisture. Much of the plant litter on the surface of soil is desiccated during dry periods, and the resulting moisture limitations to microbes may cause large accumulations of dry litter. In deserts, various arid-environment-adapted detritivores are more important as primary decomposers than they are in other ecosystems. They may include nematodes, termites, acarids, millipedes, and other arthropods (Crawford, 1979; Hadley and Szarek, 1981). Consequently, during the decomposition of surface litter, very often the soil heterotrophic microorganisms participate only as secondary decomposers following the soil fauna.

It has often been stated as self-evident that the heterotrophic microbial activities in desert soils are correlated with the available litter input, yet actual decomposition studies are scarce. Sullivan (1942) incorporated organic matter, lucerne, and grass into soil in southern Arizona. As expected, there was a high initial rate of decomposition, as measured by CO_2 evolution, decreasing with time. A further study by Lyda and Robinson (1969) was performed again with lucerne crop residue incorporated into Mojave desert soil. The soil organic matter decrease was closely correlated with the respiration rate. These studies were performed with dried green plant parts, but not with the actual litter. To use representative material, Mack (1971) collected litter from *Artemisia tridentata* and placed weighted quantities in soil under the plant canopies. Nondecomposed litter remaining was approximately 55–58% of the initial values after 30 days. A further study of similar nature was performed by Comanor and Staffeldt (1978), who collected litter from *A. tridentata* and *Larrea tridentata* and placed it *in situ* in the Great Basin and Chihuahuan Deserts, respectively. Carbon dioxide evolution from *L. tridentata* litter in the Chihuahuan site was closely related to soil moisture content. However, the CO_2 losses did not correspond totally to weight losses. Leaching probably accounted for a significant percentage of the weight loss. The final weight change was approximately 50% for *A. tridentata* after 1 year. Weight loss correlated most often with temperature and relative humidity, rather than with precipitation. The Great Basin is a cold desert, and decomposition during the winter period probably accounted for a significant weight loss. It was noted that the general downward trend of nitrogen content for many samples was interrupted by increases that the authors found difficult to explain. Heterotrophic nitrogen fixation or the presence of soil microfauna may have contributed to the apparent increase of nitrogen. The rate of litter decomposition is also a function of the quality of plant materials. Xerophytes usually contain proportionally large amounts of lignin, leading to slower decomposition rates.

6.2. Proteolysis

O'Brien (1978) studied proteolysis in Chihuahuan desert soils. The numbers of proteolytic organisms in these soils ranged from 0.15×10^6 to 1.7×10^6/g soil. At the amendment rate (1 g casein/100 g soil), the decomposition rates ranged between 52 and 93 g decomposed/week per m^2 at 10-cm depth. The rates increased 3- to 4-fold at $37°C$ incubation. However, there was no significant difference among rates at 10-, 20-, and 30-cm depths. Haddad

(1972) studied biosynthesis of protease by *Bacillus* spp. and by indigenous soil microorganisms in the Chihuahuan Desert. Pure cultures of *Bacillus* spp. and the native microbial populations of the soils synthesized protease under all environmental conditions studied, indicating that protease biosynthesis by proteolytic microorganisms was partially constitutive, since the enzymes were produced in the absence of protein amendment. When protein substrates were added to soils with low numbers of proteolytic organisms, there was a lag before proteolysis was detectable. However, once the proteolytic organisms were established, the rates of proteolysis were comparable to the rates in soils with high indigenous populations. Apparently, protein decomposition in soil is controlled primarily by environmental factors, such as moisture and temperature, rather than by the initial numbers or physiological characteristics of the native soil microorganisms (O'Brien, 1978). In the presence of 2-chloro-6 (trichloromethyl) pyridine ("N-Serve"), about 72% of the added casein was ammonified in 4 weeks in Great Basin desert soils (Skujiņš and West, 1973). Half of it was ammonified during the first week at 75 and 55% water saturation [representing ψ_w −0.03 to −0.18 megapascal (MPa)]. At −12 MPa water potential during 4 weeks, 32% of the added casein was decomposed, determined as released NH_4^+. During 5 weeks, 24% of nitrogen in *Atriplex confertifolia* litter was released as NH_4^+ at −0.03 MPa and 9% at −12 MPa water potential. O'Brien (1978) obtained similar results when he examined the Chihuahuan desert soils. Twenty-five percent of nitrogen from protein was released as NH_4^+ at −9.2 MPa water potential. Although the rates of native protein conversion to NH_4^+ from plant litter were lower than the rates of added protein (casein) amendment, it was significant that not only proteolysis but also ammonification rates were considerable at very low water potentials in desert soils.

6.3. Nitrification

Although *in vitro* nitrification potential patterns of western North American (Skujiņš and Trujillo-y-Fulgham, 1978) and Presaharan (Skujiņš, 1977b) desert soils consistently show prolonged lag periods followed by increases in NO_2^- concentrations, indicating a low potential for NO_2^- oxidation to NO_3^-, in the soils *in situ,* NO_2^- is seldom found in measurable quantities and NO_3^- is present at a few micrograms per gram. Occasionally, larger concentrations are found in surface soils or accumulations in lower horizons, representing the rainfall infiltration front or geologic accumulation.

Most of the nitrifying capability of arid soils is located in the surface layer, responding to NH_4^+ availability, and decreases sharply with depth; it is apparently absent in soil horizons not reached by precipitation. Nitrifiers may be sporadically absent, especially in very arid hot sandy soils. Although leaching of NO_3^- to deeper soil horizons may take place under certain climatic and edaphic conditions, the major mechanisms for its removal are plant uptake or denitrification. Since denitrification is dependent on the availability of organic matter, NO_3^- may occasionally accumulate in the arid organic-matter-poor soils, especially if its origin was from cyanobacterial fixation. In deserts, the nitrifying microbial population appears to be adapted to arid and warmer environments, and measurable nitrification may take place at water potentials below -1.5 MPa. A considerable effect by allelochemics of desert plant species (*Atriplex, Artemisia,* and *Ceratoides*) on the nitrification process has been demonstrated (Skujiņš, 1975), especially on the conversion of NO_2^- to NO_3^-. These results correlate with the observed effects of xeric brush canopies on nitrification (Rixon, 1971; Tiedemann and Klemmedson, 1973; Charley and West, 1977). The inhibitory mechanisms on nitrification may increase NH_4^+ accumulation in soils under plant canopies and enhance retention of NO_2^-, the primary electron acceptor for the denitrification process. Since the allelopathic effects reflect increased presence of plant litter, increased denitrification of NO_2^- may result.

6.4. Denitrification

Denitrification rates in arid soils increase with increased moisture, temperature, available organic carbon, and decrease of the available C/available N ratio in plant litter. The presence of nitrogen-poor organic matter may reduce the loss of nitrogen by increasing immobilization (Westerman and Tucker, 1978). The loss of cryptogamic crust (cyanobacteria)-fixed N_2, however, may reach over 99% of the fixed N_2 (Klubek *et al.,* 1978). Although denitrification is considered an anaerobic process, the oxygen concentration in cryptogamic crust microsites need not fall below a measurable quantity for nitrification and denitrification to take place simultaneously (Payne, 1973; Skujiņš and Klubek, 1978). A study of denitrification in a *Prosopis glandulosa*-dominated area of the Sonoran Desert showed that the mean denitrification rate under the canopies was 11.6 g N/ha per hr compared to 0.2 g N/ha per hr in the interspace soil (Virginia *et al.,* 1982). The authors ascribed the higher rates under canopies to elevated soil organic matter content and to organic matter availability in the rhizosphere. They concluded, however, that only approximately 0.5 kg N/ha may be lost by denitrification following infrequent major rainfalls. Denitrification appears instrumental in a variety of arid ecosystems as a pathway of nitrogen loss.

6.5. N_2 Fixation

Biological nitrogen input in desert soils may be by three mechanisms: (1) legume–*Rhizobium* symbiosis, (2) associative symbiosis in the rhizosphere, and (3) photoautotrophic cyanobacteria, either free-living or in association with lichens. Depending on the floral, edaphic, and climatic characteristics, one or several of the three processes may function in a desert soil.

Legumes are usually present in arid ecosystems. The legume–*Rhizobium* symbiosis has been extensively studied, but information relating to arid soils is scanty. In the desert soils of New South Wales, Australia, a dense cover of *Swainsonia* may fix up to 280 kg N/ha per year (Beadle, 1959), but dense communities of such plants are rare. It is likely that perennials of the subfamily Papilionatae may be important in certain arid ecosystems, for example, *Retama retam* and *Genista* spp. in the Presaharan Desert. Extensive communities of Fabaceae shrubs exist in North American deserts, such as *Cercidium* (paloverde) and *Prosopis* (mesquite). In Senegal and the Sudan, nitrogen content in soils under *Acacia albida* was considerably higher than under nonleguminous trees and shrubs (Jung, 1967; Radwanski and Wickens, 1967). Attempts to locate nodules on *Prosopis* spp. roots in the field, usually within the surface 1-m soil layer, have given negative results and have caused speculation that these plants do not fix nitrogen in the field, although nodulation has been achieved in pot experiments. Felker and Clark (1982) have demonstrated that nodulation and nitrogen fixation are dependent on the continuous presence of water, $\psi_w > -0.7$ MPa, and may occur at depths below 2.7 m.

Not all species of the Fabaceae family can be expected to form rhizobial associations. According to Vincent (1974), 93% of species belonging to the Papilionatae subfamily have been found to form nodules; of Mimosiodeae (e.g., *Acacia, Prosopis*), 87%; and of Caesalpinioideae (e.g., *Cercidium, Cassia*), only 23%. The lack of suitable climatic conditions and the apparent absence of appropriate *Rhizobium* frequently prevent nodule formation in legumes of arid areas (Beadle, 1964). No nodules have been found on legumes in the Mojave Desert (Garcia-Moya and McKell, 1970) or in other southwestern United States deserts (Barth and Klemmedson, 1978). In the northern Egyptian desert, nodulated native legumes are active for 30–45 days during the short growing season in winter, and *Melilotus indicus* and *Vicia peregrina* may fix up to 10.9 μg N_2/plant per hr (Hammouda, 1980).

Large areas in the semiarid western United States are inhabited by *Frankia*-nodulated nonlegumes, such as *Ceanothus* spp., *Cercocarpus* spp., *Purshia tridentata,* and *Shepherdia argentea.* The significance of *Frankia*-associated N_2 fixation in nitrogen economy in semiarid soils is not yet well understood (Klemmedson, 1979). The plant species examined have shown great variation in nodulation frequency, effectiveness of N_2 fixation, and responses to environmental stresses.

The importance of associative symbiotic N_2 fixation has also become evident for xerophytes. Surveys of arid-area flora have demonstrated that numerous species of the Asteraceae, Chenopodiaceae, Krameriaceae, Liliaceae, Poaceae, Rosaceae, Rutaceae, Solanaceae, Zygophyllaceae, and other families show positive nitrogenase activity by the presumptive acetylene reduction test in their rhizospheres and rhizoplanes (Wallace et al., 1974; Hunter et al., 1975; Snyder and Wullstein, 1973; Hammouda, 1980). N_2 fixation may be limited to short seasonal time periods having favorable temperature and moisture conditions. Besides other limiting factors, heterotrophic nitrogen fixation is dependent on the available organic carbon. In organic-carbon-poor soils, plant roots are the sites of carbon availability and channels of microbial activities in deeper soil horizons, thus also becoming the sites for N_2 fixation.

Although N_2-fixing phyllospheric microorganisms may supply nitrogen for certain plants in more mesic and humid ecosystems, the phyllospheric N_2 fixation in desert plants has not been studied. The leaf galls found on Artemisia (Farnsworth, 1975) might be one possibility for such N_2 fixation.

Surfaces of desert soils are often covered by cryptogamic crusts usually dominated by cyanobacterial flora. The cyanobacterial N_2 fixation appears to be a significant mechanism for nitrogen input in arid soils (cf. Section 4.1.1).

N_2 fixation by heterotrophic, free-living microorganisms, especially Azotobacter spp., has been widely studied in arid soils (Sasson, 1972), but is currently considered of minor importance as a mechanism for nitrogen input. The lack of available carbon, next to low water availability and high temperature, is apparently the major cause for their relatively low contribution to nitrogen input. Nitrogen input by free-living heterotrophic fixation may be more pronounced in organic-matter-enriched microenvironments, for example, in the rhizosphere (Sasson, 1972), and in decaying cryptogamic crusts, especially in the "phycosphere" (Klubek and Skujinš, 1980).

The presence of NH_4^+ limits N_2 fixation. As the amount of NH_4^+ in the crust microenvironment increases, fixation decreases, approaching zero at approximately 75 μg NH_4^+-N/g soil (Klubek and Skujinš, 1980). Allelopathic chemicals released by desert plants, either volatile or leachable from litter, may also limit N_2 fixation by cyanobacteria. The deposition of allelochemicals is localized, depending on plant species and canopy extent, and their overall effect has not been resolved. Atriplex confertifolia and Artemisia tridentata, for example, exhibit strong allelopathic effects on cyanobacteria. The canopy soils under these shrubs in the Great Basin Desert are often populated by bryophytes, and cyanobacteria are usually absent, whereas the interspace soil might have a dense cryptogamic crust. The allelopathic effects may also strongly influence the microbially mediated nutrient turnover rates and, consequently, plant successional patterns.

Reviews on aspects of nitrogen processes in semiarid soils have been recently published by Vlek et al. (1981) and Skujiņš (1981).

6.6. Other Biochemical Activities

Soil respiration, dehydrogenase activity, proteolytic activity, and ATP concentration were examined in the Great Basin, Sonoran, Chihuahuan, and Tunisian Presaharan Deserts (Table IV) (Skujiņš, 1977a,b). The activities were determined at water potentials from -7.7 to -2.8 MPa, present in soils at the time of collection. At these water potentials, all soils showed respiratory activity. Similarly, a study of Great Basin desert soils (Skujiņš, 1973) showed a correlation among dehydrogenase activity, respiration, proteolysis, and nitrification. The activities correlated well vertically with depths at the collection sites. Correlation between sites on a horizontal basis was considerably poorer. Studies on Tunisian Presaharan desert soils indicated that dehydrogenase activity was present in surface layers, and only traces of activity were found below 1 cm (Skujiņš, 1977b). In general, the activity was considerably lower than in southwestern United States deserts. Dehydrogenase activity of soil pavement increased with its apparent age. Activity in tufts (hummocks or

Table IV. Comparison of Soil Biochemical Properties with Microbial Numbers and Soil Chemical Characteristics[a]

Parameter	Areas[b]							
	1	2	3	4	5	6	7	8
Soil pH	7.7	7.7	7.4	8.2	8.3	8.3	8.6	8.6
Organic C (%)	0.34	2.47	0.80	0.30	1.12	0.49	0.16	0.15
Total N (%)	0.04	0.14	0.09	0.04	0.12	0.05	0.02	0.02
Aerobic bacteria $\times 10^6$	1.01	17.3	1.65	0.83	4.8	0.93	0.33	0.78
Fungal propagules $\times 10^4$	1.22	8.1	1.55	0.65	2.7	0.57	0.84	0.86
Proteolytic bacteria $\times 10^6$	0.35	5.6	0.21	0.28	1.1	0.28	0.14	0.04
Chitinolytic bacteria $\times 10^6$	0.09	4.0	0.19	0.09	0.52	0.18	0.023	ND
Cellulolytic bacteria $\times 10^6$	0.49	1.3	0.65	0.20	0.37	0.32	ND	ND
Respiration (μm CO_2/g per min)	7.6	19.9	12.7	9.6	19.5	11.4	12.8	8.5
Dehydrogenase activity (μg formazan/g)	27	56	198	73	172	62	25	15
ATP ($\mu g/g$)	0.034	0.017	0.447	0.037	0.203	0.102	0.022	0.011
Proteolysis (% hydrolysis)	5.5	5.4	16.3	12.9	27.4	17.7	10.3	5.7

[a]Derived from Skujiņš (1977a,b). (ND) Not determined.
[b]Areas: (1) Chihuahuan Desert, Jornada, New Mexico; (2) Chihuahuan Desert, Jornada, New Mexico (playa); (3) Sonoran Desert, Silverbell, Arizona; (4) Mohave Desert, Rock Valley, Nevada; (5) Great Basin Desert, Curley Valley, Utah; (6) Great Basin Desert, Pine Valley, Utah; (7) Presaharan Desert, Dar ez Zaoui, Tunisia; (8) Presaharan Desert, Henchir es Siane, Tunisia.

nebkas) depended on the plant species present. There was limited activity in *Aristida pungens* tufts, but higher activity under the shrubs of *Rantherium suaveolens* and *Retama retam*. High dehydrogenase activity was found in annuals in sandy areas, and the activity was significantly higher in nongrazed areas as compared to similar surface areas in grazed sites. This indicated that the dehydrogenase activity might be used as a measuring stick for the recovery process of grazed sites. The respiratory activity was somewhat lower than in the North American deserts, especially in the surface soils, and apparently reflected the availability of substrate. Proteolytic activity was also present in Tunisian desert soils, although activity was considerably lower than in North American deserts. A unique finding was that the proteolytic activity was 4-fold higher under the petrocalcic layer than in the surface soil.

7. Desertification and Soil Microbiology

Due to increased human activities on arid ecosystems, much of the global desert area is subjected to removal of distinct plant cover either by cultivation, grazing, trampling, or motor vehicle use. Removal of plant cover leads to destabilization of surface soils, subjecting them to increased aeolian erosion and often considerable dune formation. It has been demonstrated repeatedly that by controlled use of these areas, the arid and semiarid ecosystems may return to their original state and increased productivity. Most of the research in this area has been devoted to reestablishing a plant cover. Practically no information is available on organic matter formation, microbiological activities, and soil biota participation in the sand-stabilization processes. It is well known, however, that microorganisms, especially fungi and actinomycetes, participate in formation of soil structure. Although there have been many investigations of dune stabilization by higher plants, little attention has been paid to the possible role of microorganisms. Studies on microbial aggregation of sand in a dune system (Forster, 1979) showed that of the three main types of aggregates formed, microbial aggregates were more important for stabilizing sand than either root–microbial or debris–microbial aggregates. Microorganisms, in particular bacteria, may play a major role in aggregating and stabilizing sand prior to colonization by higher plants. Lynn and Cameron (1973) indicated that in disturbed areas where cryptogamic crust had existed before, the reestablishment of the algal–lichen crust is slow, although the free-living algal component rapidly establishes itself preceding the appearance of any recognizable lichen component and, consequently, limits water and wind erosion.

Organic material, such as crude oil or mulches, has been used for dune stabilization. However, we know little about the microbiological processes involved in the degradation of these materials in arid soils (Skujiņš et al., 1983).

Some initial studies comparing the soil biochemical and biological properties of soils in grazed and denuded areas with those recovering after a 4-year exclusion from grazing (Skujiņš, 1977b) indicated that in the nongrazed sites, the proteolytic organisms doubled, from 14,000 to 28,000/g soil, the chitinolytic organisms increased 10-fold, and there was an increase of lipolytic organisms and carbohydrate utilizers. Changes in actinomycete and anaerobic bacterial populations were negligible. Fungi increased from 8400 to 34,000 propagules/g soil. The increase in bacterial numbers apparently reflected the plant litter availability following increased density of vegetation. Dehydrogenase activity doubled in the surface layer and increased severalfold in the 20- to 40-cm horizon in the ungrazed area. Respiratory activity increased, especially in the surface soils, and so did ATP concentration. The content of organic carbon in the soils did not change perceptibly, but the total nitrogen content decreased in the recovered site, indicating an increase in C:N ratio in the nongrazed soils. A unique phenomenon was a decrease of fixed NH_4^+ to 0.23 g N/g in grazed from 11.4 g N/g in the nongrazed area.

One would expect that in desert areas subjected to controlled grazing or completely excluded from grazing, the biological activities in soils would increase due to increased availability of organic matter. Such studies, however, should be expanded and proper parameters selected for the determination of the fate of various soil amendments and the microbial participation in the recovery processes.

References

Abdel-Ghaffar, A. S., and Fawaz, K. M., 1977, Soil bacteria, in: *Systems Analysis of Mediterranean Desert Ecosystems of Northern Egypt* (M. A. Ayyad, ed.), Progress Report No. 3, pp. 13/1–13/31, University of Alexandria, Alexandria, Egypt.

Aldon, E. F., 1978, Endomycorrhizae enhance shrub growth and survival on mine spoils, in: *The Reclamation of Disturbed Arid Lands* (R. A. Wright, ed.), pp. 174–179, University of New Mexico Press, Albuquerque.

Alexander, M., 1976, Natural selection and the ecology of microbial adaptation in a biosphere, in: *Extreme Environments* (M. R. Heinrich, ed.), pp. 3–25, Academic Press, New York.

Allen, M. F., Smith, W. K., Moore, T. S., and Christensen, M., 1981, Comparative water relations and photosynthesis of mycorrhizal and non-mycorrhizal *Bouteloua gracilis* H. B. K. Lag ex Steud., *New Phytol.* **88:**683–693.

Bagyaraj, D. J., and Menge, J. A., 1978, Interaction between a VA-mycorrhiza and *Azotobacter* and their effects on rhizosphere microflora and plant growth, *New Phytol.* **80:**567–583.

Bailey, H. P., 1979, Semi-arid climates: Their definition and distribution, in: *Agriculture in Semi-Arid Environments* (A. E. Hall, G. H. Cannell, and H. W. Lawton, eds.), pp. 73–97, Springer-Verlag, Berlin.

Barrow, N. J., Malajczuk, N., and Shaw, T. C., 1977, A direct test of the ability of vesicular–arbuscular mycorrhiza to help plants take up fixed soil phosphate, *New Phytol.* **78:**269–276.

Barth, R. C., and Klemmedson, J. O., 1978, Shrub-induced spatial patterns of dry matter, nitrogen, and organic carbon, *Soil Sci. Soc. Am. J.* **42:**804–809.

Bauman, A. J., 1976, Desert varnish and marine ferromanganese oxide nodules: Congeneric phenomena, *Nature (London)* **259:**387–388.

Beadle, N. C. W., 1959, Some aspects of ecological research in semi-arid Australia, in: *Biogeography and Ecology in Australia* (A. T. Keast, ed.), pp. 452–460, W. Junk, The Hague, Netherlands.

Beadle, N. C. W., 1964, Nitrogen economy in arid and semiarid plant communities. III. The symbiotic nitrogen-fixing organisms, *Proc. Linn. Soc. N. S. W.* **89:**273–286.

Biederbeck, V. O., Campbell, C. A., and Nickolaichuk, W., 1977, Simulated dew formation and microbial growth in soil of a semiarid region of western Canada, *Can. J. Soil Sci.* **57:**93–102.

Binet, P., 1981, Short-term dynamics of minerals in arid ecosystems, in: *Arid Land Ecosystems: Structure, Functioning and Management,* Vol. 2 (D. W. Goodall and R. A. Perry, eds.), pp. 325–356, Cambridge University Press, Cambridge.

Bishay, A., and McGinnies, W. G. (eds.), 1979, *Advances in Desert and Arid Land Technology and Development,* Vol. 1, Harwood Academic Publishers, Chur, Switzerland.

Booth, W. E., 1941, Algae as pioneers in plant succession and their importance in erosion control, *Ecology* **22:**38–46.

Borut, W., 1960, An ecological and physiological study of soil fungi in the Northern Negev (Israel), *Bull. Res. Counc. Isr. Sect. D* **8:**65–80.

Brock, T. D., 1975, Effect of water potential on a *Microcoleus* (Cyanophyceae) from a desert crust, *J. Phycol.* **11:**316–320.

Brock, T. D., 1979, Ecology of saline lakes, in: *Strategies of Microbial Life in Extreme Environments* (M. Shilo, ed.), pp. 29–47, Verlag Chemie, Weinheim.

Brockman, E. R., 1976, Myxobacters from arid Mexican soil, *Appl. Environ. Microbiol.* **32:**642–644.

Brown, A. D., 1976, Microbial water stress, *Bacteriol. Rev.* **40:**803–846.

Brown, A. D., 1979, Physiological problems of water stress, in: *Strategies of Microbial Life in Extreme Environments* (M. Shilo, ed.), pp. 65–81, Verlag Chemie, Weinheim.

Cameron, R. E., 1969, Cold desert characteristics and problems relevant to other arid lands, in: *Arid Lands in Perspective* (W. G. McGinnies and B. J. Goldman, eds.), pp. 167–205, University of Arizona Press, Tucson.

Cameron, R. E., 1972, Microbial and ecological investigations in Victoria Valley, southern Victoria Land, Antarctica, in: *Antarctica Terrestrial Biology* (G. A. Llano, ed.), pp. 195–260, American Geophysical Union, Washington, D.C.

Cameron, R. E., and Blank, G. B., 1965, A. Soil studies: Microflora of desert regions. VIII. Distribution and abundance of desert microflora, in: *JPL Space Programs Summary No. 37-34,* Vol. IV, pp. 193–202, California Institute of Technology, Pasadena.

Cameron, R. E., Blank, G. B., Gensel, D. R., and Davies, R. W., 1965, C. Soil properties of samples from the Chile Atacama Desert, in: *JPL Space Programs Summary No. 37-35,* Vol. IV, pp. 214–223, California Institute of Technology, Pasadena.

Cameron, R. E., Honour, R. C., and Morelli, F. A., 1976, Antarctic microbiology, in: *Extreme Environments* (M. R. Heinrich, ed.), pp. 57–82, Academic Press, New York.

Charley, J. L., 1972, The role of shrubs in nutrient cycling, in: *Wildland Shrubs—Their Biology and Utilization* (C. M. McKell, J. P. Blaisdell, and J. R. Goodin, eds.), pp. 182–203, U.S. Department of Agriculture Forest Service, General Technical Report INT-1, Ogden, Utah.

Charley, J. L., and West, N. E., 1975, Plant-induced soil chemical patterns in some shrub-dominated semi-desert ecosystems in Utah, *J. Ecol.* **63:**945–963.

Charley, J. L., and West, N. E., 1977, Micropatterns of nitrogen mineralization activity in soils of some shrub-dominated semi-desert ecosystems in Utah, *Soil Biol. Biochem.* **9**:357–365.

Comanor, P. L., and Staffeldt, E. E., 1978, Decomposition of plant litter in two western North American deserts, in: *Nitrogen in Desert Ecosystems* (N. E. West and J. Skujiņš, eds.), pp. 31–49, Dowden, Hutchinson, and Ross, Stroudsburg, Pennsylvania.

Cook, R. J., and Pappendick, R. I., 1970, Soil water potential as a factor in the ecology of *Fusarium roseum* f. sp. *cerealis* "Culmorum," *Plant Soil* **32**:131–145.

Crawford, C. S., 1979, Desert detritivores: A review of life history patterns and trophic roles, *J. Arid Environ.* **2**:31–42.

Cundell, A. M., 1977, The role of microorganisms in the revegetation of strip-mined land in the western United States, *J. Range Manage.* **30**:299–305.

Davidson, R. W., and Mielke, J. L., 1947, *Fomes robustus,* a heart-rot fungus on cacti and other desert plants, *Mycologia* **39**:210–217.

Diab, A., 1978, Studies on thermophilic microorganisms in certain soils in Kuwait, *Zentralbl. Bakteriol. Parasitenkd. II* **133**:579–587.

Diab, A., and Zaidan, A., 1976, Actinomycetes in the desert of Kuwait, *Zentralbl. Bakteriol. Parasitenkd. II* **131**:545–554.

Dogan, A., 1975, Some effects of microflora on surface runoff quality, Ph.D. dissertation, Utah State University, Logan.

Dommergues, Y., 1962a, Contribution a l'étude de la dynamique microbienne des sols en zone semi-aride et zone tropicale sèche. I, *Ann. Agron.* **13**:265–324.

Dommergues, Y., 1962b, Contribution a l'étude de la dynamique microbienne des sols en zone semi-aride et zone tropicale sèche. II, *Ann. Agron.* **13**:391–468.

Dommergues, Y., and Mangenot, F., 1970, *Écologie Microbienne du Sol,* Masson et Cie, Paris.

Dorn, R. I., and Oberlander, T. M., 1981, Microbial origin of desert varnish, *Science* **213**:1245–1247.

Dregne, H. E., 1968, Appraisal of research on surface materials of desert environments, in: *Deserts of the World* (W. G. McGinnies, B. J. Goldman, and P. Paylore, eds.), pp. 287–377, University of Arizona Press, Tucson.

Dregne, H. E., 1976, *Soils of Arid Regions,* Elsevier, Amsterdam.

Dubost, D., and Hethener, P., 1966, Aperçu microbiologique des sols de deux oasis du Tassili N'Ajjer: Djanet et Iheria, *Trav. Inst. Rech. Sahariennes* **25**:7–27.

Durrell, L. W., and Shields, L. M., 1960, Fungi isolated in culture from soils of the Nevada test site, *Mycologia* **52**:636–641.

Elvidge, C. D., and Moore, C. B., 1979, A model for desert varnish formation, *Geol. Soc. Am. Abstr. Program* **11**:271.

Elwan, S. H., and Diab, A., 1970a, Studies in desert microbiology. II. Development of bacteria in the rhizosphere and soil of *Artemisia monosperma* Del. in relation to environment, *U. A. R. J. Bot.* **13**:97–108.

Elwan, S. H., and Diab, A., 1970b, Studies in desert microbiology. III. Certain aspects of the rhizosphere effect of *Rhazya stricta* Decn. in relation to environment, *U. A. R. J. Bot.* **13**:109–119.

Elwan, S. H., and Diab, A., 1970c, Studies in desert microbiology. IV. Bacteriology of the root region of a fodder xerophyte in relation to environment, *U. A. R. J. Bot.* **13**:159–169.

Elwan, S. H., and Diab, A., 1970d, Studies in desert microbiology. V. Certain patterns of bacterial development in relation to depth and environment, *U. A. R. J. Bot.* **13**:171–179.

Elwan, S. H., and Diab, A., 1976, Actinomycetes of an Arabian desert soil, *Egypt. J. Bot.* **19**:111–114.

Elwan, S. H., and Mahmoud, S. A. Z., 1960, Note on the bacterial flora of the Egyptian desert in summer, *Arch. Microbiol.* **36**:360–364.

Englund, B., 1975, Potential nitrogen fixation by blue-green algae in some Tunisian and Swedish soils, *Plant Soil* **43**:419–431.

Evenson, A. E., 1961, A preliminary report of the myxomycetes of southern Arizona, *Mycologia* **13**:137–144.

Farnsworth, R. B., 1975, Nodulation and nitrogen fixation in shrubs, in: *Proceedings of the Symposium and Workshop on Wildland Shrubs* (H. C. Stutz, ed.), pp. 32–71, Brigham Young University Press, Provo, Utah.

Faust, W. F., 1971, Blue-green algal effects on some hydrologic processes at the soil surface, in: *Hydrology and Water Resources in Arizona and the Southwest, Proceedings of the 1971 Meeting of the Arizona Section of the American Water Resources Association, and the Hydrological Section of the Arizona Academy of Sciences,* pp. 99–105, Tempe, Arizona.

Felker, P., and Clark, P. R., 1982, Position of mesquite (*Prosopis* spp.) nodulation and nitrogen fixation (acetylene reduction) in 3-m long phraetophytically simulated soil columns, *Plant Soil* **64**:297–305.

Fletcher, J. E., 1960, Some effects of plant growth on infiltration in the southwest, in: *Water Yield in Relation to Environment in the Southwestern United States,* American Association for the Advancement of Science Symposium, pp. 51–63, Houston, Texas.

Fletcher, J. E., and Martin, W. P., 1948, Some effects of algae and molds in the rain-crust of desert soils, *Ecology* **29**:95–100.

Focht, D. D., and Martin, J. P., 1979, Microbiological and biochemical aspects of semiarid agricultural soils, in: *Agriculture in Semi-Arid Environments* (A. E. Hall, G. H. Cannell, and H. W. Lawton, eds.), pp. 119–147, Springer-Verlag, Berlin.

Forster, S. M., 1979, Microbial aggregation of sand in an embryo dune system, *Soil Biol. Biochem.* **11**:537–543.

Franz, G., 1971, Mikrobiologische Characterisierung einiger natürlicher und kultivierter Standorte in drei verschiedenen ökologischen Regionen Chiles, *Plant Soil* **34**:133–158.

Friedmann, E. I., 1982, Endolithic microorganisms in the antarctic cold desert, *Science* **215**:1045–1053.

Friedmann, E. I., and Galun, M., 1974, Desert algae, lichens, and fungi, in: *Desert Biology* (G. E. Brown, ed.), pp. 165–212, Academic Press, New York.

Friedmann, E. I., and Ocampo, R., 1976, Endolithic blue-green algae in the dry valleys: Primary producers in the antarctic desert ecosystem, *Science* **193**:1247–1249.

Friedmann, E. I., Lipkin, Y., and Ocampo-Paus, R., 1967, Desert algae of the Negev (Israel), *Phycologia* **6**:185–196.

Fuller, W. H., 1974, Desert soils, in: *Desert Biology,* Vol. 2 (G. E. Brown, ed.), pp. 31–101, Academic Press, New York.

Garcia-Moya, E., and McKell, C. M., 1970, Contribution of shrubs to the nitrogen economy of a desert-wash plant community, *Ecology* **51**:81–88.

Gerdemann, J. W., and Trappe, J. M., 1974, The Endogonaceae in the Pacific Northwest, *Mycologia Mem.* **5**:1–76.

Gifford, G. F., 1972, Infiltration rate and sediment production trends on a plowed big sagebrush site, *J. Range Manage.* **25**:53–55.

Grenot, C. J., 1974, Physical and vegetational aspects of the Sahara desert, in: *Desert Biology,* Vol. 2 (G. E. Brown, ed.), pp. 103–164, Academic Press, New York.

Griffin, D. M., 1972, *The Ecology of Soil Fungi,* Chapman and Hall, London.

Griffin, D. M., 1981, Water and microbial stress, in: *Advances in Microbial Ecology,* Vol. 5 (M. Alexander, ed.), pp. 91–136, Plenum Press, New York.

Griffin, D. M., and Luard, E. J., 1979, Water stress and microbial ecology, in: *Strategies of Microbial Life in Extreme Environments* (M. Shilo, ed.), pp. 49–63, Verlag Chemie, Weinheim.

Haddad, S. G., 1972, Identification of *Bacillus* sp. and the ecological significance of its protease, M.S. thesis, New Mexico State University, Las Cruces.

Hadley, N. F., and Szarek, S. R., 1981, Productivity of desert ecosystems, *BioScience* **31**:747–753.

Hall, I. R., 1978, Effects of endomycorrhizas on the competitive ability of white clover, *N. Z. J. Agric. Res.* **21**:509–515.

Hall, I. R., and Armstrong, P., 1979, Effect of vesicular–arbuscular mycorrhizas on growth of white clover, lotus, and ryegrass in some eroded soils, *N. Z. J. Agric. Res.* **22**:479–484.

Hammouda, F. M. M., 1980, Biological nitrogen fixation in the desert ecosystems of Egypt, Ph.D. thesis, Alexandria University, Alexandria, Egypt.

Hanks, R. J., and Ashcroft, G. L., 1980, *Applied Soil Physics—Soil Water and Temperature Applications,* Springer-Verlag, Berlin.

Heinrich, M. R. (ed.), 1976, *Extreme Environments,* Academic Press, New York.

Hethener, P., 1967, Activité microbiologique des sols à *Cupressus dupreziana* A. Camus au Tassili N'Ajjer (Sahara central), *Bull. Soc. Hist. Nat. Afr. Nord* **58**:39–100.

Hirrell, M. C., Mehravaran, H., and Gerdemann, J. W., 1978, Vesicular–arbuscular mycorrhizae in the Chenopodiaceae and Cruciferae: Do they occur?, *Can. J. Bot.* **56**:2813–2817.

Horowitz, N. H., 1979, Biological water requirements, in: *Strategies of Microbial Life in Extreme Environments* (M. Shilo, ed.), pp. 15–27, Verlag Chemie, Weinheim.

Hunt, C. B., and Durrell, L. W., 1966, Distribution of fungi and algae, *U.S. Geol. Surv. Prof. Pap.* **509**:55–66.

Hunter, R. B., Wallace, A., Romney, E. M., and Wieland, P. A. T., 1975, Nitrogen transformations in Rock Valley and adjacent areas of the Mohave Desert, *US/IBP Desert Biome Res. Memo. 75-35,* Utah State University, Logan.

Johnson, R. M., 1973, Microbial investigations at the Chaabania site in Tunisia, in: *Systems Analysis of the Presaharan Ecosystem of Southern Tunisia,* Report No. 1 (G. Novikoff, F. H. Wagner, and D. Skouri, eds.), pp. 46–57, Utah State University, Logan.

Jung, G., 1967, Influence de l'*Acacia albida* (Del.) sur la biologie des sols dior, ORSTOM, Dakar, Senegal.

Khudairi, A. K., 1969, Mycorrhiza in desert soils, *BioScience* **19**:598–599.

Killian, C., and Feher, D., 1935, Recherches sur les phénomènes microbiologiques des sols sahariens, *Ann. Inst. Pasteur* **55**:573–623.

Killian, C., and Feher, D., 1938, Le rôle et l'importance de l'exploration microbiologique des sols sahariens, in: *La Vie dans la Region Desertique Nord-tropicale de l'Ancien Monde,* pp. 81–106, Société de Bio-géographie, Paris.

Killian, C., and Feher, D., 1939, Recherches sur la microbiologie des sols désertiques, *Encycl. Biol. (Paris)* **21**:1–23.

Klemmedson, J. O., 1979, Ecological importance of actinomycete-nodulated plants in the western United States, *Bot. Gaz.* **140**(Suppl.):S91–S96.

Klubek, B., and Skujiņš, J., 1980, Heterotrophic N_2-fixation in arid soil crusts, *Soil Biol. Biochem.* **12**:229–236.

Klubek, B., Eberhardt, P. J., and Skujiņš, J., 1978, Ammonia volatilization from Great Basin desert soils, in: *Nitrogen in Desert Ecosystems* (N. E. West and J. Skujiņš, eds.), pp. 107–129, Dowden, Hutchinson, and Ross, Stroudsburg, Pennsylvania.

Knight, W. G., and Skujiņš, J., 1981, ATP concentration and soil respiration at reduced water potentials in arid soils, *Soil Sci. Soc. Am. J.* **45**:657–660.

Kovda, V. A., Samoilova, E. M., Charley, J. L., and Skujiņš, J., 1979, Soil processes in arid lands, in: *Arid-land Ecosystems: Structure, Functioning and Management,* Vol. 1 (D. W. Goodall, R. A. Perry, and K. M. W. Howes, eds.), pp. 439–470, Cambridge University Press, Cambridge.

Kushner, D. J., 1978, Life in high salt and solute concentrations: Halophilic bacteria, in: *Microbial Life in Extreme Environments* (D. J. Kushner, ed.), pp. 317–368, Academic Press, London.

Lange, O. L., Schulze, E. D., and Koch, W., 1970a, Experimentell-ökologische Untersuchungen an Flechten in Negev-Wüste. II. CO_2-Gaswechsel und Wasserhaushalt von *Ramalina maciformis* (Del.) Bory am natürlichen Standort während der sommerlichen Trockenperiod, *Flora (Jena) Abt. B* **159**:38–62.

Lange, O. L., Schulze, E. D., and Koch, W., 1970b, Experimentell-ökologische Untersuchungen an Flechten in Negev-Wüste. III. CO_2-Gaswechsel und Wasserhaushalt von Krusten- und Blattflechten am natürlichen Standort während der sommerlichen Trockenperiode, *Flora (Jena) Abt. B* **159**:525–528.

Lanyi, J. K. (reporter), 1979, Life at low water activities, in: *Strategies of Microbial Life in Extreme Environments* (M. Shilo, ed.), pp. 125–135, Verlag Chemie, Weinheim.

Larkin, J. M., and Dunigan, E., 1973, Myxobacteria from southwestern USA soils, *Soil Sci. Soc. Am. Proc.* **37**:808–809.

Lindsey, D. L., Cress, W. A., and Aldon, E. G., 1977, The effects of endomycorrhizae on growth of rabbitbrush, fourwing saltbush, and corn in coal mine spoil material, USDA Forest Service Research Note RM-343.

Lipman, C. B., 1912, The distribution and activities of bacteria in soils of the arid regions, *Univ. Calif. Publ. Agric. Sci.* **1**:1–20.

Llano, G. A. (ed.), 1972, *Antarctic Terrestrial Biology,* American Geophysical Union, Washington, D.C.

Lobova, E. V., 1960, *Pochvy pustynnoi zony SSSR,* Akademiya Nauk, Moscow (translation: Lobova, E. V., 1967, *Soils of the Desert Zone of the USSR,* U.S. Department of Commerce, Springfield, Virginia).

Long, H., and Miller, V. M., 1945, A new desert *Coprinus, Mycologia* **37**:120–123.

Lyda, S. D., and Robinson, G. D., 1969, Soil respiratory activity and organic matter depletion in an arid Nevada soil, *Soil Sci. Soc. Am. Proc.* **33**:92–94.

Lynn, R. I., and Cameron, R. E., 1973, The role of algae in crust formation and nitrogen cycling in desert soils, *US/IBS Desert Biome Research Memorandum 73-40,* Utah State University, Logan.

MacGregor, A. N., and Johnson, D. E., 1971, Capacity of desert algal crusts to fix atmospheric nitrogen, *Soil Sci. Soc. Am. Proc.* **35**:843–844.

Mack, R. N., 1971, Mineral cycling in *Artemisia tridentata,* Ph.D. dissertation, Washington State University, Pullman.

Mahendrappa, M. K., Smith, R. L., and Christiansen, A. T., 1966, Nitrifying organisms affected by climatic region in western United States, *Soil Sci. Soc. Am. Proc.* **30**:60–62.

Mahmoud, S. A. Z., Abou-El-Fadl, M., and El-Mofty, M. K., 1964, Studies on the rhizosphere microflora of a desert plant, *Folia Microbiol.* **9**:1–8.

Marks, C. G., and Kozlowski, T. T. (eds.), 1973, *Ectomycorrhizae—Their Ecology and Physiology,* Academic Press, New York.

Mayland, H. F., and McIntosh, T. H., 1966, Availability of biologically fixed atmospheric nitrogen-15 to higher plants, *Nature (London)* **209**:421–422.

McLaren, A. D., and Skujiņš, J., 1967, The physical environment of microorganisms in soil, in: *The Ecology of Soil Bacteria* (T. R. G. Gray and D. Parkinson, eds.), pp. 3–24, Liverpool University Press, Liverpool.

Meigs, P., 1953, World distribution of arid and semi-arid homoclimates, *Reviews of Research on Arid Zone Hydrology, UNESCO Arid Zone Programme* **1**:203–209.

Meigs, P., 1957, Arid and semi-arid climatic types of the world, *Proceedings, 17th International Geographical Congress, 1952, 8th General Assembly,* pp. 135–138.

Miller, J. T., and Brown, I. C., 1938, Observations regarding soils of northern and central Mexico, *Soil Sci.* **46**:427–451.

Miller, R. M., 1979, Some occurrences of vesicular–arbuscular mycorrhiza in natural and disturbed ecosystems of the Red Desert, *Can. J. Bot.* **57**:619–623.

Montasir, A. H., Mostafa, M. A., and Elwan, S. H., 1958a, Desert rhizospheric microflora as a biotic factor in the development of *Farsetia aegyptiaca*. I. Development and cultural studies of fungi and bacteria in *Farsetia* rhizosphere, *A'in Shams Sci. Bull.* **3**:83–98.

Montasir, A. H., Mostafa, M. A., and Elwan, S. H., 1958b, Desert rhizospheric microflora as a biotic factor in the development of *Farsetia aegyptiaca*. II. Effect of *Farsetia* root metabolites on conidial germination and mycelial growth of *Aspergillus terreus* and colonial growth of *Bacillus subtilis*, *A'in Shams Sci. Bull.* **3**:99–110.

Montasir, A. H., Mostafa, M. A., and Elwan, S. H., 1959a, Desert rhizospheric microflora as a biotic factor in the development of *Farsetia aegyptiaca*. III. Germination potentialities of *Farsetia* seeds in response to *Aspergillus terreus* and *Bacillus subtilis* metabolites obtained at different environmental and physiological conditions, *A'in Shams Sci. Bull.* **4**:1–8.

Montasir, A. H., Mostafa, M. A., and Elwan, S. H., 1959b, Desert rhizospheric microflora as a biotic factor in the development of *Farsetia aegyptiaca*. IV. Effect of *Aspergillus terreus* and *Bacillus subtilis* metabolites obtained at varied physiological conditions on the relative vigour of *Farsetia* seedlings, *A'in Shams Sci. Bull.* **4**:10–24.

Montasir, A. H., Mostafa, M. A., and Elwan, S. H., 1959c, Desert rhizospheric microflora as a biotic factor in the development of *Farsetia aegyptiaca*. V. Nitrogen fixation and decomposition processes by *Farsetia* rhizospheric micro-organisms, *A'in Shams Sci. Bull.* **4**:27–49.

Mosse, B., Stribley, D. P., and LeTacon, F., 1981, Ecology of mycorrhizae and mycorrhizal fungi, in: *Advances in Microbial Ecology,* Vol. 5 (M. Alexander, ed.), pp. 137–210, Plenum Press, New York.

Mouchacca, J., and Joly, P., 1974, Étude de la mycoflore des sols arides de l'Égypte. I. Le genre *Penicillium, Rev. Écol. Biol. Sol* **11**:67–88.

Mouchacca, J., and Joly, P., 1976, Étude de la mycoflore des sols arides de l'Égypte. II. Le genre *Aspergillus, Rev. Écol. Biol. Sol* **13**:293–313.

Naguib, A. I., and Mouchacca, J. S., 1971, The mycoflora of Egyptian desert soils, *Bull. Inst. Égypte* **52**:37–61.

Nicolson, T. H., and Johnston, C., 1979, Mycorrhiza in the Gramineae. 3. *Glomus fasciculatus* as the endophyte of pioneer grasses in a maritime sand dune, *Trans. Br. Mycol. Soc.* **72**:261–268.

Nicot, J., 1955, Remarques sur les peuplements de micromycètes des sables désertiques, *C. R. Acad. Sci.* **240**:2082–2084.

Nicot, J., 1960, Some characteristics of the microflora in desert sands, in: *The Ecology of Soil Fungi* (D. Parkinson and J. S. Waid, eds.), pp. 94–97, Liverpool University Press, Liverpool.

Oberholzer, P. C. J., 1936, The decomposition of organic matter in relation to fertility in arid and semi-arid regions, *Soil Sci.* **41**:359–379.

O'Brien, R. T., 1978, Proteolysis and ammonification in desert soils, in: *Nitrogen in Desert Ecosystems* (N. E. West and J. Skujiņš, eds.), pp. 50–59, Dowden, Hutchinson, and Ross, Stroudsburg, Pennsylvania.

Payne, W. J., 1973, Reduction of nitrogenous oxides by microorganisms, *Bacteriol. Rev.* **37**:409–452.

Pochon, J., de Barjac, H., and Lajudie, J., 1957, Recherches sur la microflore des sols sahariens, *Ann. Inst. Pasteur* **92**:833–836.

Poma, E., 1955, La simbiosi micorrizica nelle piante annue, *Allionia* **2**:429–442.

Price, R. S., 1911, The roots of some North African desert grasses, *New Phytol.* **10**:328–339.

Radwanski, S. A., and Wickens, G. E., 1967, The ecology of *Acacia albida* on mantle soils in Zalingei, Jebbel Marra, Sudan, *J. Appl. Ecol.* **4**:569–579.

Ranzoni, F. V., 1968, Fungi isolated in culture from soils of the Sonoran desert, *Mycologia* **60**:356–371.

Rayss, T., 1959, Champignons hypogés dans les régions désertiques d'Israel, in: *Omagiu lui Traian Săvulescu*, pp. 655–659, Rumanian Academy of Sciences, Bucharest.

Rayss, T., and Borut, S., 1958, Contributions to the knowledge of soil fungi in Israel, *Mycopathol. Mycol. Appl.* **10**:142–174.

Reeves, F. B., Wagner, D., Moorman, T., and Kiel, J., 1979, The role of endomycorrhizae in revegetation practices in the semiarid West. 1. A comparison of incidence of mycorrhizae in severely disturbed vs. natural environments, *Am. J. Bot.* **66**:6–13.

Reichenbach, H., 1970, A new myxobacterium of the family Sorangaceae, *Arch. Microbiol.* **70**:119–138.

Renaut, J., and Sasson, A., 1970, Les cyanophycees du Maroc: Etude preliminaire de quelques biotopes de la region de Rabat, *Bull. Soc. Sci. Nat. Phys. Maroc.* **50**:37–52.

Rivkind, L., 1929, Étude des terres du Sahara, *Arch. Inst. Pasteur Algérie* **7**:88–103.

Rixon, A. J., 1971, Oxygen uptake and nitrification by soil within a grazed *Atriplex vesicaria* community in semi-arid rangeland, *J. Range Manage.* **24**:435–439.

Rougieux, R., 1963, Actions antibiotiques et stimulantes de la truffe du désert (*Terfezia boudieri* Chatin), *Ann. Inst. Pasteur* **105**:315–318.

Rougieux, R., 1966, Contribution à l'étude de l'activité microbienne en sol désertique (Sahara), Thesis, University of Bordeaux.

Rychert, R. C., and Skujiņš, J., 1974a, Nitrogen fixation by blue-green algae-lichen crusts in the Great Basin Desert, *Soil Sci. Soc. Am. Proc.* **38**:768–771.

Rychert, R. C., and Skujiņš, J., 1974b, Inhibition of algal-lichen crust nitrogen fixation in desert shrub communities, *Abstr. Am. Soc. Microbiol. Annu. Meet.*, Chicago, **1974**:E20.

Rychert, R. C., Skujiņš, J., Sorensen, D., and Porcella, D., 1978, Nitrogen fixation by lichens and free-living microorganisms in deserts, in: *Nitrogen in Desert Ecosystems* (N. E. West and J. Skujiņš, eds.), pp. 20–30, Dowden, Hutchinson, and Ross, Stroudsburg, Pennsylvania.

Sanders, F. E., Mosse, B., and Tinker, P. B. (eds.), 1975, *Endomycorrhizas*, Academic Press, New York.

Sasson, A., 1967, Recherches ecophysiologiques sur la flore bacterienne de sols de regions arides du Maroc, Thesis, University of Paris.

Sasson, A., 1972, Microbial life in arid environments: Prospects and achievements, *Ann. Arid Zone* **2**:67–91.

Shilo, M. (ed.), 1979, *Strategies of Microbial Life in Extreme Environments*, Verlag Chemie, Weinheim.

Skujiņš, J., 1973, Dehydrogenase: An indicator of biological activities in arid soils, *Bull. Ecol. Res. Comm. (Stockholm)* **17**:235–241.

Skujiņš, J., 1975, Nitrogen dynamics in stands dominated by some major cool desert shrubs. IV. Inhibition by plant litter and limiting factors of several processes, *US/IBP Desert Biome Res. Memo 75-33*, Utah State University, Logan.

Skujiņš, J., 1977a, Comparison of biological processes in western deserts, *US/IBP Desert Biome Res. Memo 77-20*, Utah State University, Logan.

Skujiņš, J., 1977b, Soil microbiological and biochemical investigations, 1972–1976, in: *Systems Analysis of the Presaharan Ecosystem of Southern Tunisia*, Report No. 6 (G. Novikoff, F. H. Wagner, and M. S. Hajjaj, eds.), pp. 215–249, Utah State University, Logan.

Skujiņš, J., 1981, Nitrogen cycling in arid ecosystems, *Ecol. Bull. (Stockholm)* **33**:477–491.

Skujiņš, J., and Klubek, B., 1978, Nitrogen fixation and cycling by blue-green algae–lichen crusts in arid rangeland soils, *Ecol. Bull. (Stockholm)* **26**:164–171.

Skujiņš, J., and Trujillo-y-Fulgham, P., 1978, Nitrification in Great Basin desert soils, in: *Nitrogen in Desert Ecosystems* (N. E. West and J. Skujiņš, eds.), pp. 60–74, Dowden, Hutchinson, and Ross, Stroudsburg, Pennsylvania.

Skujiņš, J., and West, N. E., 1973, Nitrogen dynamics in stands dominated by some major cold desert shrubs, *US/IBP Desert Biome Res. Memo. 73–35,* Utah State University, Logan.

Skujiņš, J., McDonald, S. O., and Knight, W. G., 1983, Metal ion availability during biodegradation of waste oil in semiarid soils, *Ecol. Bull. (Stockholm)* **35**:341–350.

Snyder, J. M., and Wullstein, L. H., 1973, The role of desert cryptogams in nitrogen fixation, *Am. Midl. Nat.* **90**:257–265.

Sorensen, D. L., and Porcella, D. B., 1974, Nitrogen erosion and fixation in cool desert soil–algal crusts in northern Utah, *US/IBP Desert Biome Res. Memo. 74-37,* Utah State University, Logan, Utah.

Staffeldt, E. E., and Vogt, K. B., 1975, Mycorrhizae of desert plants, *US/IBP Desert Biome Res. Memo. 75-37,* Utah State University, Logan.

Staley, J. T., Palmer, F., and Adams, J. B., 1982, Microcolonial fungi: Common inhabitants of desert rocks?, *Science* **215**:1093–1095.

Stewart, W. D. P., 1967, Transfer of biologically fixed nitrogen in a sand slack region, *Nature (London)* **214**:603–604.

Sullivan, M. J., Jr., 1942, The effect of field applications of organic matter on the properties of some Arizona soils, Ph.D. dissertation, University of Arizona, Tucson.

Thornthwaite, C. W., 1948, An approach toward a rational classification of climate, *Geogr. Rev.* **38**:55–94.

Thornton, H. G., 1953, Some problems presented by the microbiology of arid soils, in: *Desert Research, Proceedings of the International Symposium,* Jerusalem, May 1952, Special Publication No. 2, pp. 295–300, Research Council of Israel, Jerusalem.

Tiedemann, A. R., and Klemmedson, J. O., 1973, Effect of mesquite on physical and chemical properties of the soil, *J. Range Manage.* **26**:27–29.

Trappe, J. M., 1981, Mycorrhizae and productivity of arid and semiarid rangelands, in: *Advances in Food Producing Systems for Arid and Semiarid Lands* (J. Manassah and E. J. Briskey, eds.), pp. 581–599, Academic Press, New York.

Vargues, H., 1953, Etude microbiologique de quelques sols sahariens en relation avec la présence d'*Anabasis aretioides* Coss et Moq, in: *Desert Research, Proceedings of the International Symposium,* Jerusalem, May 1952, Special Publication No. 2, pp. 318–324, Research Council of Israel, Jerusalem.

Vincent, J. M., 1974, Root-nodule symbioses with *Rhizobium,* in: *The Biology of Nitrogen Fixation* (A. Quispel, ed.), pp. 266–307, North-Holland, Amsterdam.

Virginia, R. A., Jarrell, W. M., and Franco-Vizcaino, E., 1982, Direct measurement of denitrification in a *Prosopis* (mesquite) dominated Sonoran desert ecosystem, *Oecologia* **53**:120–122.

Vlek, P. L. G., Fillery, I. R. P., and Burford, J. R., 1981, Accession, transformation, and loss of nitrogen in soils of the arid region, *Plant Soil* **58**:133–175.

Vogel, S., 1955, Niedere "Fensterpflanzen" in der südafrikanischen Wüste, *Beitr. Biol. Pflanz.* **31**:45–135.

Vollmer, A. T., Au, F., and Bamberg, S. A., 1977, Observations on the distribution of microorganisms in desert soil, *Great Basin Nat.* **37**:81–86.

Wallace, A., and Romney, E. M., 1972, Approximate age of shrub clumps at the Nevada Test Site, in: *Radioecology and Ecophysiology of Desert Plants at the Nevada Test Site* (A. Wallace and E. M. Romney, eds.), pp. 307–309, U.S. Atomic Energy Commission, Office of Information Services, Springfield, Virginia.

Wallace, A., Romney, E. M., Cha, J. W., and Soufi, S. M., 1974, Nitrogen transformations in Rock Valley and adjacent areas of the Mojave Desert, *US/IBP Desert Biome Res. Memo. 74-36,* Utah State University, Logan.

Wallace, A., Romney, E. M., and Hunter, R. B., 1978, Nitrogen cycle in the northern Mojave Desert: Implications and predictions, in: *Nitrogen in Desert Ecosystems* (N. E. West and J. Skujiņš, eds.), pp. 207–218, Dowden, Hutchinson, and Ross, Stroudsburg, Pennsylvania.

Went, F. W., and Stark, N., 1968, The biological and mechanical role of soil fungi, *Proc. Natl. Acad. Sci. U.S.A.* **60:**497–509.

Westerman, R. L., and Tucker, T. C., 1978, Denitrification in desert soils, in: *Nitrogen in Desert Ecosystems* (N. E. West and J. Skujiņš, eds.), pp. 75–106, Dowden, Hutchinson, and Ross, Stroudsburg, Pennsylvania.

Williams, S. E., and Aldon, E. G., 1976, Endomycorrhizal (vesicular–arbuscular) associations of some arid zone shrubs, *Southwest Nat.* **20:**437–444.

Williams, W. D. (ed.), 1981, *Salt Lakes,* W. Junk, The Hague.

Wullstein, L. H., 1980, Nitrogen fixation associated with rhizosheaths of Indian ricegrass used in the stabilization of the Slick Rock, Colorado, tailings pile, *J. Range Manage.* **33:**204–206.

Wullstein, L. H., and Pratt, S. A., 1981, Scanning electron microscopy of rhizosheaths of *Oryzopis hymenoides, Am. J. Bot.* **68:**408–419.

Wullstein, L. H., Bruening, M. L., and Bollen, W. B., 1979, Nitrogen fixation associated with sand grain root sheaths (rhizosheaths) of certain xeric grasses, *Physiol. Plant.* **48:**1–4.

3

The Input and Mineralization of Organic Carbon in Anaerobic Aquatic Sediments

DAVID B. NEDWELL

1. Introduction

The bottom sediments in both marine and freshwater ecosystems are important sites of mineralization and nutrient recycling, particularly where there is shallow water together with high productivity so that there is rapid input of organic carbon to the sediment. In most coastal and intertidal areas and in eutrophic lakes, productivity is relatively high and detrital input to the bottom sediments is appreciable, with the result that much of the sediments in these regions is anaerobic and reduced, apart perhaps from a thin aerobic surface layer. Therefore, at least potentially, a considerable portion of the organic carbon mineralization in these aquatic ecosystems may go on in the sediment under anaerobic rather than aerobic conditions.

Although some anaerobic microbial communities such as the rumen or anaerobic sewage digesters have been studied for a considerable time because of their obvious economic or practical importance, the extensive anaerobic microbenthic community has attracted systematic and intensive investigation only recently, over perhaps the last ten years. This enhanced interest has resulted both from an increased realization of the importance of the benthic communities in carbon mineralization and nutrient recycling in aquatic systems and from a rapidly intensifying interest on the part of microbiologists in

DAVID B. NEDWELL ● Department of Biology, University of Essex, Colchester C04 3SQ, Essex, United Kingdom.

the way that anaerobic microbial communities function. The classic work of Bryant *et al.* (1967) in elucidating the nature of the *"Methanobacillus omelianskii"* consortium and the subsequent enunciation of the concept of interspecies H_2 transfer illustrated how an understanding of the interactions between different functional parts of an anaerobic microbial community were vital to understanding overall community activity—in other words, the activity of the whole community was more than the simple sum of the isolated individual activities. The recent advances in understanding of anaerobic sedimentary communities have been possible only because of the complementary nature of concurrent laboratory and field studies. Isolation and description of new types of anaerobic bacteria (for example, by Widdel, 1980) and the description in the laboratory of a variety of hitherto unknown syntrophic and competitive interactions between anaerobic bacteria (Bryant *et al.*, 1967; Ferry and Wolfe, 1976; Abram and Nedwell, 1978a; McInerney *et al.*, 1979) have provided a conceptual framework in which both to design an experimental approach to the further examination of anaerobic communities and to interpret the data obtained. Again, the development of an experimental approach to the examination of anaerobic sedimentary microbial communities, and of the interactions among different parts of the community, owes much to the pioneering work of Cappenberg and co-workers with the anaerobic sediments of Lake Vechten, Holland. The stimulus that their work provided to the examination of the microbial communities in a variety of anaerobic aquatic sediments has led to at least an initial appreciation of the processes involved in carbon flow in these environments.

2. Origins of Detritus

Bottom sediments are major sites of detrital mineralization and nutrient recycling in aquatic systems, although the amount of detritus that enters a sediment, and its composition, may vary dramatically. Organic detritus has been defined (Fenchel and Blackburn, 1979) as "the organic carbon lost by non-predatory processes from any trophic level," and these processes will include ingestion, excretion, secretion, death, and decay. Detritus in bodies of water may arise directly or indirectly from the biomass produced by phytoplankton primary production, by macrophyte primary production, or by detritus brought in by runoff from the land. The proportionate contribution of each of these sources will obviously vary depending on whether freshwater, estuarine, coastal seawater, or deep oceanic sites are considered. In small eutrophic lakes where the littoral zone represents a relatively large proportion of the surface area, a large proportion of the detritus may be derived from macrophytes. For example, in Lawrence Lake, Michigan, it was calculated (Wetzel *et al.,*

1972) that macrophytes contributed 51% of the net primary production compared to only 25% by the phytoplankton and 22% by epiphytic algae. Input of allochthonous carbon, mainly in runoff during the spring period, was equivalent to about 10% of the *in situ* primary production, but this proportion will vary from lake to lake depending on the size of the catchment area from which runoff to the lake occurs. In coastal marine environments, macrophytes may also contribute significantly to the detrital input into the bottom sediments (Webster *et al.,* 1975), and in addition, detrital input from land runoff may be appreciable (Stephens *et al.,* 1967; Qasim and Sankaranarayanan, 1972; Ansell, 1974). In both the Cochin Backwater, India (Qasim and Sankaranarayanan, 1972), and off Vancouver Island, Canada (Stephens *et al.,* 1967), it was estimated that about half the detrital input to the bottom sediment was allochthonous material. With increasing distance from land, however, the contribution of phytoplankton to primary production increases, and in the majority of the deep ocean, away from terrestrial influence, virtually all detritus will be derived from this source. Excretion by both living plants and animals may also contribute to detritus, and it has been shown that 10–30% of algal primary production may be excreted as soluble compounds, particularly glycollate (Fogg, 1966; Khailov and Burlakova, 1969). Wetzel *et al.* (1972) calculated that in Lawrence Lake, about 6% of annual phytoplankton primary production was excreted.

Detritus is present both as particulate organic material (POM) and dissolved organic material (DOM). In most natural water, there is about an order of magnitude more DOM than POM (Parsons, 1963; Wetzel *et al.,* 1972; Fenchel and Blackburn, 1979). It has been pointed out (Fenchel and Blackburn, 1979) that these categories are arbitrary, since there is a continuum from dissolved organic molecules, through extremely small particles, to particles of macroscopic size. However, the division into POM and DOM is a convenient functional one, usually based on whether or not particles are filtered out by 0.2 or 0.45 μm pore size filters. There are a variety of interactions between DOM and POM during decomposition of detritus (Fig. 1), and it is only in the dissolved state that organic molecules can be taken up into their cells by microorganisms and metabolized.

3. Inputs of Detritus to Bottom Sediments

In addition to variation in the sources of detritus from site to site, the total amount of detritus and its chemical composition will vary. Primary biomass is subjected to grazing as well as direct conversion to detritus. In coastal and estuarine ecosystems where macrophyte productivity is high, only a small proportion of the primary biomass may be grazed (Mann, 1976), although in

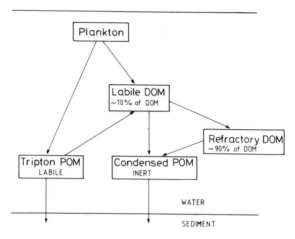

Figure 1. Conversions between particulate organic material (POM) and dissolved organic material (DOM) in a water column and input of POM to sediment. Adapted from Degens and Mopper (1976).

planktonic communities the grazing food chain may predominate (Steele, 1974). Detritus will therefore be contributed to by both the primary producers and the consumers that utilize the primary biomass. In the deep ocean, where the water column may be many thousands of meters in depth, virtually all the primary biomass is consumed by zooplankton. Fecal pellets of grazing copepods may be a significant component of the particulate detritus in both freshwater and marine ecosystems (Jones, 1976; Honjo, 1978). Honjo (1978) showed that in the water column of the Sargasso Sea, two types of fecal pellets could be distinguished—green pellets, which were produced in the euphotic zone containing photosynthetic pigments and undissolved algal cells, and red pellets, containing little photosynthetic pigment and containing coccolith cells showing dissolution. It was suggested that the latter type were formed from green pellets reingested during settlement. The fecal pellets were colonized on their surfaces by bacteria during their settlement through the water column and were subjected to decay from the outside inward (Honjo and Roman, 1978).

Although the dissolved organic material (DOM) initially derived from dead cells contains labile material, the majority (about 90%) of DOM in water is refractory (Degens and Mopper, 1976). This reflects the rapid biological removal of the labile portion. Godshalk and Wetzel (1978a) differentiated the soluble component of aquatic angiosperm detritus into four molecular-weight categories (<30,000, <10,000, <1000, and <500 daltons). Fluorescence and UV absorbance measurements during the course of microbial degradation of these soluble fractions showed that low-molecular-weight compounds with unsaturated bonds or ring structures were rapidly removed. The amount of high-molecular-weight compounds also decreased somewhat, but proportionately less than the low-molecular-weight fraction, and there was accumulation of refractory humic compounds with time. Thus, there was a proportional increase in the high-molecular-weight refractory component of the remaining DOM. This corresponds to data that suggest that the bulk of DOM in the oceanic euphotic zone is of molecular weight less than 5000, whereas at depths greater than 5000 m, most DOM is of molecular weight greater than 10,000 and is very resistant to microbial attack (Barber, 1968; Degens and Mopper, 1976). Therefore, DOM should be subdivided into "young labile" and "old refractory" components (Fig. 1). It has been suggested that cross-linking resulting from complexation with metal ions and condensation reactions between DOM molecules enhances the resistance to microbial attack of otherwise labile molecules such as amino acids and sugars (Degens and Mopper, 1976; Mopper, 1980). It has also been demonstrated that DOM may condense, particularly at a gas–liquid interface, to form particulate organic material (POM) (Menzel, 1966).

During settlement through a deep water column, POM is continuously ingested, digested, and mineralized until only the refractory components of both plant and animal debris may remain to reach the bottom sediment. Honjo (1978) showed that at greater than 5000 m depth in the Sargasso Sea, only 10% of the total particles were larger than 62 μm, and most of these were the skeletal remains of plankton together with clay particles. These considerations suggest that in deep water columns, only a small proportion of the biomass will survive mineralization to be deposited on the bottom sediment, but the proportion will tend to be greater in the comparatively shallow waters of lakes and coastal seas (Suess and Muller, 1980). For example, it was calculated that in the Sargasso Sea, only 2% of the organic carbon formed during primary production in the water column reached the bottom at 5300 m (Honjo, 1978), compared to 27% in the 30-m water column of Loch Thurnaig, Scotland (Davies, 1975), or 43% in the 14.5-m water column of Kaneohe Bay, Hawaii (Taguchi, 1982). Again, Pennington (1974) calculated that in the deep oligotrophic lakes Ennerdale and Wastwater in the English Lake District, there was 50% loss of sedimenting seston carbon, compared to only 25% loss in the shallower eutrophic Blelham Tarn. Since total productivity is greater in coastal and estuarine compared to deep oceanic seawater, and in eutrophic compared to oligotrophic lakes, the total amounts of organic carbon settling to the sediments will also reflect these relative productivities (Suess and Muller, 1980). Table I (see also Lastein, 1976) shows values derived from a number of studies illustrating organic carbon fluxes into bottom sediments over a range of water-column depths.

David B. Nedwell

Table I. Some Estimates of Annual Inputs of Organic Carbon to Bottom Sediments in a Variety of Aquatic Ecosystems

Ecosystem	Water depth (m)	Sedimentation (g C/m^2 per year)	Authority
Freshwater			
Ennerdale, England, oligotrophic	42	29	Pennington (1974)
Wastwater, England, oligotrophic	76	28	Pennington (1974)
Blelham Tarn, England, eutrophic	14.5	158	Pennington (1974)
Lawrence Lake, U.S.A., eutrophic	5	28	Wetzel et al. (1972)
Lawrence Lake, U.S.A., eutrophic	10.5	33[a]	Wetzel et al. (1972)
Horw Bay, Lake Lucerne, Switzerland, eutrophic	60	58	Bloesch et al. (1977)
Rotsee, mesotrophic	14	142	Bloesch et al. (1977)
Marine			
St. Margaret's Bay, Nova Scotia	10	134	Webster et al. (1975)
King Edward's Cove, South Georgia Island	11–18	60	Platt (1979)
Kaneohe Bay, Hawaii	14.5	155	Taguchi (1982)
Echernforde Bight, Baltic Sea	20	25	Iturriaga (1979)
Loch Thurnaig, Scotland	20–30	28	Davies (1975)
Baltic Sea	25	40	Zeitschel (1965) (in Webster et al., 1975)
Off Vancouver Island, Canada	30	200	Stephens et al. (1967)
St. Margaret's Bay, Nova Scotia	70	118	Webster et al. (1975)
Tongue of the Ocean, Bahamas	2000	2.1	Wiebe et al. (1976)
Continental slope off NE U.S.A.	2200–3650	0.04	Rowe and Gardner (1979)
Sargasso Sea	5563	0.46	Honjo (1978)

[a]The greater amount depositing at 10.5 m was attributed to transport of POM from the shallows into the deep part of the lake.

Since the settlement time of particulate material will be related to the depth of the water column, decomposition during settlement tends to be less in shallow water, and detritus entering the sediment will therefore also tend to contain a greater remaining proportion of labile components (Berner, 1978, 1980b; Molongoski and Klug, 1980a). It has been demonstrated that the organic carbon content of the superficial layers of bottom sediments generally decreases with increased water depth (Vigneaux et al., 1980) and is dependent on both primary production in the water column and the rate of sedimentation (Suess and Muller, 1980). Toth and Lerman (1977) concluded from relative changes of the C:N:P ratios in the surface layers of benthic sediments that the reactivity of the sedimentary organic material was related to the rate of deposition, the material being more reactive and more rapidly mineralized where sedimentation rates were high.

In those freshwater and marine aquatic systems where the water column is comparatively shallow, the input of organic carbon to the sediment may change seasonally, reflecting seasonal changes in plankton or macrophyte productivity or in the amount of allochthonous material from land runoff (Stephens *et al.*, 1967; Wetzel *et al.*, 1972; Pennington, 1974; Davies, 1975). Seasonal peaks of organic carbon deposition onto bottom sediments may also reflect the disappearance of stratification in the water column and the release and settlement of POM previously trapped at the metalimnion (Jones, 1976; Lastein, 1976).

4. Organic Carbon in Sediments

Settling organic detritus will become incorporated into the surface layers of sediment by further deposition or by mixing of the surface layers of sediment by bioturbation (Hylleberg and Henriksen, 1980; Fry, 1982). Organic detritus within the sediment will be utilized by benthic organisms and will, with time, be mineralized. Increased depth in a sediment also represents increased time since deposition and vertical profiles through the sediment should show a decrease of organic material with depth, at least below the surface mixed layer. It has been pointed out (Billen, 1982) that the initial breakdown of dissolved organic material (DOM) into soluble smaller units that can be taken up and metabolized by microorganisms is brought about by exoenzymatic hydrolysis (see Burns, 1978), and this initial hydrolysis is probably the rate-limiting step for degradation of detrital organic carbon.

4.1. Diagenetic Models

The decay of organic material in anaerobic sediments has been modeled by diagenetic analysis (Jørgensen, 1978; Berner, 1980a,b; Billen, 1982). The organic detritus deposited onto the sediment is regarded as being composed of a number of categories (G_i) of organic material of varying degrees of lability to microbial attack (see Section 6.2.1). The rate of decay (R_i) of any category can be modeled by first-order kinetics (Berner, 1980a; Billen, 1982):

$$R_i = k_i \cdot G_i$$

where G_i is the concentration of class i of organic compounds and k_i is the first-order decay constant for G_i.

The total organic material (G) within the sediment is the sum of all categories:

$$G = \sum_i G_i$$

and the overall rate of organic matter degradation (R) is

$$R = \sum_i k_i \cdot G_i$$

Each class of organic material will decrease exponentially with time (that is, with depth within the sediment) at rates that are dependent on its lability or refractility. Thus, the labile part of the deposited organic carbon will tend to be rapidly removed (high k_i) within the surface layers of sediment, whereas refractory components will be only slowly degraded (low k_i) and persist for long periods of time (that is, to great depth). Some detrital components, such as lignins under anaerobic conditions (Zeikus, 1981), may not be susceptible to microbial attack at all $(k = 0)$ and will therefore persist and accumulate within the sediment.

Berner (1980b) diagenetically modeled organic matter degradation within the anaerobic region of a Long Island Sound sediment where organic carbon mineralization was largely by sulfate-reducing bacteria. He used a "2G" model, where G_α was labile particulate nondiffusible material decomposed within the zone of bioturbation and a fraction G_β decomposed only below the bioturbated layer. With the use of this model, theoretical profiles of residual organic carbon concentration with depth were calculated. The shape of the calculated organic carbon concentration profile with depth was similar to that actually measured *in situ*, but the metabolizable carbon $(G_\alpha + G_\beta)$ considered in the model represented only about half the measured total organic carbon. The difference was attributed to the presence of highly refractory organic geopolymers that were formed in the 1-cm-deep aerobic layer at the sediment surface.

Billen (1982) has extended this analysis with a "3G" model to data (Benoit *et al.*, 1979; Aller and Yingst, 1980; Aller *et al.*, 1980; Turekian *et al.*, 1980) for the same muddy sediment site in Long Island Sound. In the surface bioturbated zone, organic carbon decreased rapidly with depth as labile, easily used organic material (G_1) was removed. A second class (G_2) of more refractile organic material was only slowly decomposed, while a third class (G_3), which was possibly of terrestrial origin (Benoit *et al.*, 1979), was almost entirely refractory to microbial degradation within the sedimentary environment. This model fitted well to the organic carbon profiles measured *in situ* (Fig. 2) and illustrates that changes of organic material with sedimentary depth can be suc-

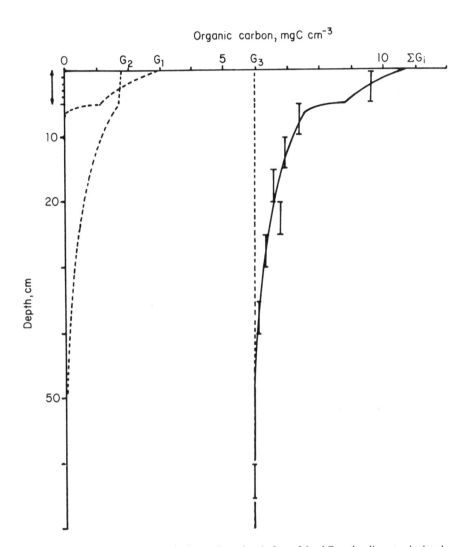

Figure 2. Vertical profiles of residual organic carbon in Long Island Sound sediment calculated with a "$3G$" diagenetic model. The distributions of three categories of organic material (G_1, labile; G_2, slowly metabolized; G_3, refractory) are shown, together with the calculated total residual organic carbon content (————). ($\mathbf{I}$) Measured *in situ* concentrations of organic carbon; ($\updownarrow$) extent of the bioturbated layer. From Billen (1982).

cessfully modeled using only a restricted number of categories of organic material.

4.2. Measurements of Microbially Labile Carbon in Sediments

The work of Seki *et al.* (1968) has shown that the amount of soluble microbiologically labile organic material present in the surface layer of coastal marine sediments may vary seasonally, the amount in the winter sediment from Departure Bay, Canada, being greater than in summer sediment. This presumably reflected inhibition of microbial activity in the winter sediment by low winter temperature leading to accumulation of labile carbon. The decrease in labile dissolved organic matter with increased depth has been similarly investigated in saltmarsh sediment. Figure 3 shows the total organic carbon content of the anaerobic sediment from a pan in the Colne Point saltmarsh, Essex, United Kingdom, as measured with a Total Carbon analyzer (Beckman Instruments Inc., Fullerton, California), together with the amount of soluble, available, DOM in the pore water. Total organic carbon increased with depth as the sediment was compacted, reflecting the accumulation of organic carbon resistant to anaerobic microbial attack. The amount of available DOM in the pore water of the sediment was measured with a microbiological assay method similar to that of Seki *et al.* (1968), with *Pseudomonas aeruginosa* as the assay organism. Growth of *Ps. aeruginosa* was proportional to the amount of available carbon in the pore water included in the assay medium and was expressed in terms of equivalent growth on glucose. The amount of soluble available carbon in pore water decreased sharply with depth, indicating that labile organic carbon was present in the pore water only within the top 2 cm of sediment.

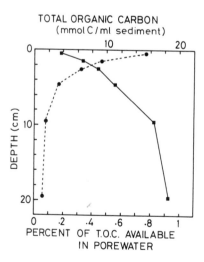

Figure 3. Vertical profiles of total organic carbon (T.O.C.) and microbially available organic carbon in the pore water of the anaerobic sediment of a saltmarsh pan. T.O.C. was measured with a carbon analyzer and the microbially available carbon in the pore water by a bioassay using *Pseudomonas aeruginosa*. (■) T.O.C.; (●) % of T.O.C. that is microbially available.

Similarly, other data have shown that turnover of fatty acids (Balba and Ned-well, 1982) and mineralization of organic nitrogen (Abd. Aziz and Nedwell, 1979) are detectable only within the top 5 cm of the sediment profile. These observations emphasize that labile organic carbon, easily available for micro-bial utilization, is confined to the surface layer. Although the total organic car-bon content of the sediment increased with depth, the proportion that was available for microbial metabolism decreased from about 1% at the surface to $< 0.1\%$ at 20 cm depth, although at least part of this difference reflects a decrease in the pore-water content of the sediment at 20 cm due to compaction.

5. Electron Acceptors for Organic Carbon Oxidation

5.1. Extent of the Surface Aerobic Layer

The mineralizing activity of microorganisms in bottom sediments depends on the availability of labile organic carbon. In aerobic surface layers of sedi-ment aerobic respiratory metabolism will predominate; therefore, the overall importance of aerobic metabolism to the total amount of organic carbon miner-alization within a sediment will be a function of the extent of the aerobic layer. This extent will in turn depend on the balance between transport of oxygen into the sediment from the overlying oxygenated water column and its removal by respiration within the sediment. Meio- and macrobenthic fauna may con-tribute to benthic respiration, but the respiration of the microbenthos is prob-ably the single most important aerobic respiratory process within most aquatic sediments (Fry, 1982). However, benthic invertebrates may significantly increase the transport rate of oxygen into the surface layers of sediment above that due to diffusion alone both by irrigation of the sediment resulting from respiratory currents of water pumped through their burrows and by physical overturn and mixing (bioturbation) of the surface layer by their feeding and burrowing activities (Hylleberg and Henriksen, 1980; Fry, 1982). The result of such activity is to extend the depth of the oxidized surface layer and thereby to enhance aerobic mineralization. This is important, since the overall rate of mineralization of organic carbon tends to be greater aerobically than anaero-bically (e.g., Godshalk and Wetzel, 1978a; Jørgensen, 1980).

Environmental factors that increase the rate of aerobic respiration will, in contrast, diminish the penetration of oxygen into a sediment (Bouldin, 1968; Howeller, 1972). Thus, lack of bioturbation, increased temperature (which will stimulate microbial respiration rates), high organic content (which results in a more abundant supply of electron donors and hence higher respiration rates), and small particle size of sediments (which decreases the diffusion rates of solutes) will all tend to limit oxygen penetration into a sediment and diminish the extent of the aerobic surface layer.

Dissolved oxygen concentrations will decrease more or less rapidly with increased sedimentary depth, depending on the balance between oxygen influx from the sediment surface and its removal within the sediment. The aerobic layer may vary from many meters of depth in the sediments of the deep ocean, where both organic content and temperature are low, to only a few millimeters of depth in highly organic freshwater, intertidal, or coastal marine sediments during the summer (Billen and Verbeustel, 1980; Revsbech et al., 1980). Indeed, where respiratory removal of oxygen by the sedimentary microflora exceeds the rate of influx of oxygen, the water above the sediment may itself become deoxygenated. It is the oxygen demand of the bottom sediments that creates anaerobic hypolimnia in many freshwater and marine situations (Jones, 1976).

5.2. Sequential Use of Electron Acceptors with Depth

Removal of oxygen within the sediment is accompanied by chemical reduction of the sediment, as shown by the typical decrease in the redox potential (Eh) of sediment with increased depth. However, the extent of the oxygenated surface layer may not coincide with the extent of the chemically oxidized layer as defined by $Eh > 0$ mv. Dissolved oxygen may be entirely removed at depths considerably less than those at which the sediment becomes chemically reduced (Revsbech et al., 1980). As oxygen becomes depleted, and redox conditions suitably reduced, other electron acceptors are utilized by different physiological types of anaerobic bacteria. The sequence of use of these electron acceptors can be predicted on thermodynamic grounds according to the amount of energy that will be liberated from a common electron donor as it is oxidized by each of the electron acceptors (Mechalas, 1974; Billen and Verbeustel, 1980; Billen, 1982). Thermodynamic criteria suggest the sequence O_2, Mn^{4+}, NO_3^-, Fe^{3+}, SO_4^{2-}, HCO_3^- as the electron acceptors become less oxidizing and less energy is available. Figure 4 illustrates the approximate range of Eh over which each electron acceptor is microbially metabolized and the products of its reduction. This sequence may, however, be modified by other nonthermodynamic criteria such as competition for electron donors or variations in the availability or concentration of the various electron acceptors in different environments. However, other things being equal, there will tend to be a vertical sequence with depth of successive utilization by bacteria of the electron acceptors in the order suggested above as the Eh of the sediment becomes progressively more reduced with increased depth. Examination of a variety of sedimentary environments tends to provide data supporting this vertical distribution sequence (Froelich et al., 1979; Jørgensen, 1980), although the sequence may be locally modified.

Apart from the aerobes, any of these physiological groups of bacteria, respiring a particular electron acceptor, depends on the activity of the preced-

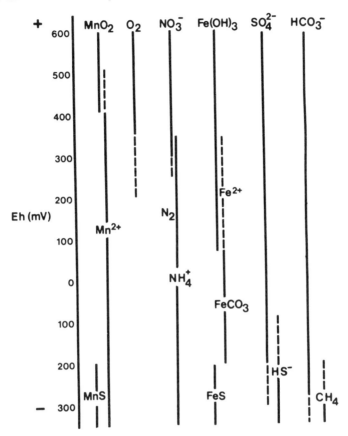

Figure 4. Approximate Eh ranges for microbial utilization of potential electron acceptors. From Jones (1982).

ing group(s) to reduce the environmental Eh or oxygen tension or both to a level at which they are themselves able to commence metabolism. Many of these anaerobic bacteria are inhibited or killed in oxidizing environments or by oxygen in their environment (Morris, 1975). In turn, their own activity may change the environmental Eh to a value at which succeeding microorganisms find the environment amenable and are able to commence metabolism. Thus, the total mineralization within a vertical column of sediment will be the sum of mineralization of organic carbon substrates by the respiratory activity of all these groups of aerobic and anaerobic microorganisms, together with microbial fermentation reactions, which can also lead to mineralization of organic carbon to CO_2. Fermentation reactions, in the absence of O_2, are not redox-dependent and will occur throughout the anaerobic part of the sedimentary profile.

5.3. Importance of Different Electron Acceptors in Marine and Freshwater Sediments

The importance of any one electron acceptor to overall organic carbon mineralization may vary markedly, depending at least partly on differences in the availability of each electron acceptor in different sedimentary environments. A comparison of two sediments, one marine and one freshwater, illustrates the differences in the amounts of organic carbon calculated to be oxidized by each of oxygen, nitrate, sulfate, or carbon dioxide (Table II). In the marine Limfjord sediment sulfate reduction accounted for over half the organic carbon oxidized, but was unimportant in the freshwater Blelham Tarn sediment where dissolved sulfate concentrations were low. Nitrate reduction was relatively more important in the eutrophic Blelham Tarn sediment than in the Limfjord, where seawater contained little nitrate. The use of CO_2 as an electron acceptor by methanogenic bacteria is inhibited in the marine sediment by the presence of active sulfate reduction (Abram and Nedwell, 1978a) (see also Section 6.5), but in the low-sulfate freshwater sediment CO_2 reduction to CH_4 was much more important. Indeed, in the anaerobic profundal sediments of some North American lakes as much as 50–60% of the deposited organic carbon may be mineralized to CH_4 in the absence of other major electron acceptors such as O_2, NO_3^-, or SO_4^{2-} (Fallon et al., 1980). The relative importance of the different electron acceptors may also vary among different areas within a lake. After allowing for the different areas of littoral and profundal sediments, Jones and Simon (1981) estimated that sulfate reduction and methanogenesis in Blelham Tarn were relatively more important to carbon mineralization in the anaerobic profundal sediment, whereas nitrate reduction and aerobic oxidation of carbon were more important within the less reduced littoral sediments.

Jørgensen (1982) has recently emphasized the importance of sulfate-reducing bacteria to organic carbon oxidation in marine sediments of the continental shelf. He calculated that in shallow-water sediments, with high organic input and community O_2 uptake rates greater than 20–25 mmoles O_2/m^2 per

Table II. Oxidation of Organic Carbon by Different Electron Acceptors in a Freshwater and a Marine Sediment

Sediment	Percentage of total organic carbon oxidized by:				Reference
	O_2	NO_3^-	SO_4^{2-}	HCO_3^-	
Freshwater					
Blelham Tarn, U.K.	42	17	2	25	Jones and Simon (1980)
Marine					
Limfjord, Denmark	46	3	51	0	Jørgensen (1980)

day, sulfate reduction accounted for 50% of the organic carbon oxidation and aerobic metabolism for the other 50%. In deeper-water sediments, with a smaller organic input, sulfate reduction accounted for only 35% and aerobic metabolism for 65% of the organic carbon oxidation. This proportionate decrease in the importance of sulfate vs. carbon oxidation was attributed to the more extensive aerobic layer in these deep-water sediments, where aerobic microbial oxidation could mineralize a greater proportion of the organic carbon input before sulfate reduction became possible in the anaerobic region. The majority of global sulfate reduction (>90%) appears to occur in the organically rich shallow water (<200 m depth) marine sediments, where organic input is high and the surface aerobic layer is restricted in depth.

6. Processes in Detrital Decay and Organic Carbon Mineralization in Anaerobic Sediments

6.1. Structure of the Anaerobic Microbial Community

In aerobic sedimentary communities, detritus is degraded and oxidized by a complex community of microorganisms including bacteria, fungi, and protozoa together with larger meio- and macrofauna. The interactions of these components of the biological community may be extremely important to detrital decay. For example, ingestion and maceration of detritus within the intestinal tracts of meiofauna increase its surface area and render it more vulnerable to subsequent microbial attack (Hargrave, 1972). Again, grazing of bacteria by protozoa may stimulate the rate of detrital breakdown (Fenchel and Harrison, 1976). The interactions contributing to detrital breakdown have been the subject of a number of reviews and books (Fenchel and Harrison, 1976; Saunders, 1976; Fenchel and Jørgensen, 1977; Fenchel and Blackburn, 1979; Lee, 1980), and the reader is referred to these for further details. A common feature of all aerobic organisms is that they are capable of bringing about the complete mineralization of organic molecules to CO_2 via the tricarboxylic acid cycle, and all the organisms involved in the aerobic processing of detritus therefore contribute directly to mineralization of organic carbon. This is not the case in anaerobic environments, where mineralization requires the interaction of a complex microbial community, each part of the community contributing a partial oxidation of organic carbon until finally complete mineralization is achieved. A further feature is that in the terminal stages of anaerobic carbon flow organic carbon may be mineralized either by its complete oxidation to CO_2 or by its reduction to CH_4.

Anaerobic microbial communities have been differentiated into a number of functionally different groups of bacteria (Wolfe and Higgins, 1979), each of which contributes to a different stage of anaerobic carbon mineralization. The

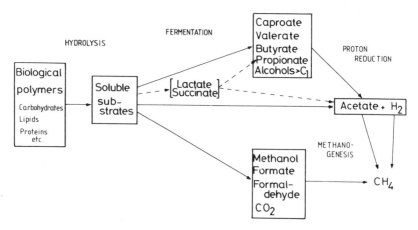

Figure 5. A generalized illustration of the pathways of carbon flow, and the functional groups of bacteria, in an anaerobic sedimentary microbial community.

community is divided (Fig. 5) into (1) hydrolytic microorganisms that bring about the initial breakdown of polymeric material and the fermentation of hydrolysis products to intermediary fermentation products such as fatty acids and alcohols; (2) acetogenic microorganisms that further metabolize intermediary fermentation products to yield mainly acetate, H_2, and CO_2; and (3) microorganisms involved in the terminal steps of organic carbon oxidation that utilize the products of earlier stages, predominantly H_2 and acetate. In reality, the divisions may not be well defined in all cases. For example, some bacteria are able to metabolize carbohydrates directly to acetate + H_2 in the presence of an H_2-scavenging bacterium without the intermediate accumulation of fermentative products such as fatty acids or alcohols (Iannotti *et al.*, 1973; Winter and Wolfe, 1980). Such microorganisms are therefore hydrolyzers, but are also acetogenic. Carbon flow in these cases by-passes the pools of intermediary fermentative products and is oxidized directly to acetate + H_2. Which of these two alternative general pathways to acetate predominates will be important in determining the relative sizes of the pools of intermediary organic molecules that are found in the sediment. This model of the anaerobic community, however, provides a convenient basis for discussion of the effect of microbial activity on organic carbon flow in sediments.

6.2. Initial Breakdown

Those organisms that are responsible for the initial hydrolysis and fermentation of polymeric material present within the sediment, often in particulate form, produce extracellular or membrane-bound hydrolytic enzymes (Burns, 1978) that bring about the hydrolysis of polymers such as cellulose, starch, hemicellulose, proteins, and chitin to their soluble monomers. These soluble monomers can then be metabolized fermentatively to yield intermediary products including fatty acids, alcohols, CO_2, and H_2. Because soluble organic molecules are intermediary products of the initial anaerobic breakdown of detrital material, rather than CO_2 as under aerobic conditions, the production of dissolved organic material (DOM) is usually greater during anaerobic than during aerobic detrital decay (Otsuki and Hanya, 1972a,b).

6.2.1. Differential Decay Rates of Detrital Components

Detritus is derived from the death and subsequent breakdown of living biomass. Cells contain a complex mixture of cytoplasmic components ranging from biologically labile ones such as sugars and amino acids to very refractory ones such as lignins, chitin, and waxes that are relatively resistant to microbiological attack (Hackett et al., 1977; Maccubbin and Hodson, 1980; Zeikus, 1981). These different components of detritus will decay at different relative rates depending on their lability. Studies of detrital decomposition have revealed typically exponential courses of decay with time (Otsuki and Hanya, 1972a,b; Fallon and Pfaender, 1976; Godshalk and Wetzel, 1978b,c). Initially, fast rates of weight loss and mineralization of the detrital material are observed as the biologically labile components of the detritus are rapidly removed. However, increasing proportions of the remaining detritus are resistant refractory molecules that are degraded only slowly, and the rate of decay therefore diminishes. For example, in an examination of degradation by lake bacteria of detritus from five aquatic angiosperm species, Godshalk and Wetzel (1978b) showed that weight loss was described by first-order kinetics:

$$dW/dt = -kW$$

where W is the percentage remaining of initial detritus weight, t is time, and k is the first-order decay rate constant.

Rate constants varied between 0.002 and 0.085/day depending on temperature and whether the detritus was incubated aerobically or anaerobically. After about 50 days, this simple model did not fit well to the data, and a better fit was obtained if it was assumed that k also decreased exponentially with time:

$$k = ae^{-bt}$$

where a and b are decay constants, and e is the base of natural logarithms.
Overall

$$dW/dt = a(e^{-bt})W$$

These same workers examined the rates of disappearance of different components of the detritus and showed that soluble nonstructural carbohydrates decreased rapidly, but fibrous structural carbohydrates (including lignin, hemicellulose, and cellulose) were degraded only slowly. The relative decay rates of the five angiosperms were inversely related to fiber content, but this was not a simple relationship, since *Zostera marina* had a similar fiber content but decayed much more slowly (Godshalk and Wetzel, 1978a). It was suggested that ultrastructural adaptation of *Zostera*'s structural components to harsh environmental conditions rendered them more resistant to microbial attack. Similarly, the lignin of *Pinus elliottii* was mineralized by saltmarsh sediment bacteria 10 times more slowly than that of *Spartina alterniflora* (Maccubbin and Hodson, 1980), and this was attributed to structural differences between the lignins of grasses and gymnosperms. The proportion of structural polymers is smaller in the cells of algae than in those of higher plants, but even so, exponential decay rates were observed during both aerobic and anaerobic decomposition of dead green algal cells at $20°C$ (Otsuki and Hanya, 1972a,b). Over the first 60 days, about 50% of the cell carbon was lost anaerobically (decay constant $0.0088/day$), 30% being converted to DOM and 20% mineralized, but the rate of decay subsequently diminished. Again, it was the algal cell walls that were most resistant to microbial attack, although the refractory structural polymers were pectins, alginates, and cellulose, rather than lignins.

Variation in the relative rates of degradation of different detrital fractions is well illustrated by the study of mineralization of ^{14}C-labeled lignocellulose in *Spartina alterniflora* detritus by saltmarsh sediment microflora (Maccubbin and Hodson, 1980). After 720 hr of aerobic incubation, 3 times more of the ^{14}C-labeled cellulose moiety had been mineralized than of the $[^{14}C]$lignin moiety (32.1 and 10.6%, respectively, of the initial content of the detritus). There was no measurable mineralization of lignin under anaerobic conditions, an observation supported by other workers (Hackett *et al.,* 1977; Zeikus, 1981), although cellulose was mineralized anaerobically.

Differences in the rates of mineralization of detrital compounds are also illustrated by changes in C:N:P ratios during detrital decomposition. Phytoplankton have C:N:P ratios near $100:16:1$ (Redfield *et al.,* 1963; Holm-Hansen, 1972; Jones, 1976). Examination of C:N:P ratios in residual detritus during decomposition has shown that in the early stages, organic nitrogen is mineralized more rapidly than organic carbon, resulting in an initial increase of the C:N ratio (Seki *et al.,* 1968; Otsuki and Hanya, 1972a; Holm-Hansen, 1972), although it may subsequently decrease again because of microbial colonization of the remaining detrital particles (Odum and de la Cruz, 1967; de la Cruz, 1975; Saunders, 1976). From changes of the C:N:P ratios of particulate organic material between the epilimnion and hypolimnion of Blelham Tarn, Jones (1976) concluded that the relative rates of mineralization during settlement of seston were in the order $N > C > P$.

Godshalk and Wetzel (1978a) have differentiated three phases in the breakdown of detritus: Phase A, in which there is rapid weight loss due to leaching and autolytic release of DOM, much of which is labile and easily metabolized by microorganisms. For example, Fallon and Pfaender (1976) studied the microbial metabolism of seawater leachates of *Spartina alterniflora* by the microbial community in saltmarsh water and found that 68% of the initial DOM was mineralized after 2 days' incubation at 15°C. In phase B, after leaching, decay rates are lower, and any weight loss is the result of degradative microbial activity. It is in this phase that microbial activity is the major cause of weight loss and also environmental variables have greatest influence on rates of decay. In Phase C, only highly refractory polymers remain, and the rate of breakdown approaches zero.

6.2.2. Polymer Hydrolysis and Fermentation

The distributions of bacterial hydrolyzers of cellulose starch, chitin, and other polymeric materials have been examined by plate counting in both marine and freshwater sediments (Zobell, 1946; Oppenheimer, 1960), although the data are sparse. As with "total" bacterial counts, there is a general trend of highest counts near the sediment–water interface, where detrital input occurs and where decay rates are greatest, and below which counts decrease. Jones (1971) calculated that of the estimated viable count of heterotrophic bacteria in the sediments of Lakes Windermere and Esthwaite, amylase-producing bacteria represented, respectively, 9.7 and 24.0%; protease-producing bacteria, 7.5 and 6.0%; and lipase-producing bacteria, 2.2 and 3.5%. The majority of exoenzyme-producing bacteria were gram-negative. Measured activities of amylase, protease, lipase, and glucosidase in the surface layer of sediment were several orders of magnitude greater than in the water column, and the rates of exoenzyme activity correlated with the counts of exoenzyme-producing bacteria. In another study, Molongoski and Klug (1976) anaerobically isolated bacteria from the anaerobic profundal sediment of Wintergreen Lake on media containing either glucose or an amino acid mixture as carbon source. Isolates were dominated by proteolytic *Clostridium* spp., *Streptococcus* spp., and *Eubacterium* spp., all of which gave fatty acid metabolic products corresponding to those typically found within the sediment. The isolates showed no cellulolytic capacity, but some showed lipase and amylase activity. Schink and Zeikus (1982) also found a *Clostridium* sp. to be the major pectinolytic anaerobe isolated from anaerobic sediment from Lake Mendota and Kuaack Lake. Pectin was mineralized to CH_4 and CO_2. The turnover to CH_4 (100 hr) was more rapid than cellulose turnover (Nelson and Zeikus, 1974), but slower than for glucose (12 hr).

Mopper (1980) has shown with acid hydrolysates of seawater that glucose is the predominant sugar, with relatively large concentrations of fructose also present, perhaps arising by isomerization from glucose. In pore waters of sediments, the free sugars are again dominated by glucose and fructose, although at concentrations as much as an order of magnitude greater than in seawater. Fructose is often more abundant than glucose in sediments, possibly because it is more strongly bound into metal complexes and further isomerization of glucose to fructose then occurs. Treatment of sediments with EDTA to release complexed carbohydrates, followed by acid hydrolysis, showed that in aerobic sediments, about 60–70% of bound carbohydrates was associated with metals compared to only 30–40% in anaerobic sediments (Mopper, 1980). Complexation and binding may reduce the accessibility of such molecules to microbial uptake and degradation (Degens and Mopper, 1976).

There are few measurements of *in situ* rates of carbohydrate turnover in anaerobic sediments, but Christian and Wiebe (1978) measured potential turnover rates of $[^{14}C]$glucose in anaerobic slurry of sediment from a Georgia saltmarsh. At two sites, the rates of glucose turnover decreased with increased depth in the sediment, and the turnover was faster in the high-sulfate sediment from the tall-*Spartina* zone (turnover times 1.3–4.2 hr) than in the low-sulfate sediment from the short-*Spartina* zone (turnover times 4–47 hr). The amount of label detected in ether extracts of intermediary products of glucose metabolism was always much lower in the high-sulfate sediment, suggesting that carbon turnover to CO_2 was more rapid than in the low-sulfate sediment.

Similarly, there are only a few direct measurements of amino acid turnover. Hanson and Gardner (1978) measured the turnover of $[^{14}C]$alanine and $[^{14}C]$aspartic acid either to CO_2 or assimilated into cell material in slurries of sediment from a Georgia saltmarsh. Dissolved free alanine and aspartate concentrations ranged from 1 to 500 nmoles/liter. As with glucose (Christian and Wiebe, 1978), turnover rates of alanine and aspartate were greater in the tall-*Spartina* zone (turnover times 5–25 hr) than in the short-*Spartina* zone (turnover times 40–100 hr). Mineralization of both alanine and aspartate decreased with depth in the tall-*Spartina* zone, but showed little change with depth in the short-*Spartina* zone. However, Hall *et al.* (1972) found that turnover rates of $[^{14}C]$glycine were 10 times faster in sediment slurries than in intact, undisturbed sediment, and slurry techniques may therefore overestimate *in situ* rates of activity.

Christensen and Blackburn (1980) measured turnover of $[^{14}C]$alanine in small cores of intact sediment from the Danish Limfjord. The measured pool size of free alanine in the sediment was 800 nmoles/liter. Biological turnover rate constants (0.11/min) did not vary significantly with depth over the top 0–10 cm, but 30% of the alanine that disappeared from the free amino acid pool was strongly adsorbed within the sediment and was only slowly metabolized to CO_2 over an extended period of time. Label from the $[^{14}C]$alanine appeared quickly in the sedimentary pool of volatile fatty acids, deamination of the amino acid being the first stage in its conversion to fatty acids. Only about 20% of the alanine was assimilated into cellular material, compared to 70–87% assimilation of alanine in saltmarsh sediment (Hanson and Gardner, 1978). Christensen and Blackburn (1980) suggested that the different assimilation efficiencies might reflect differing ratios of inorganic nitrogen to available organic carbon (energy) within the two sediments.

6.3. Exchange between Free and Adsorbed Pools of Substrates

The data of Christensen and Blackburn (1980) emphasize the difficulty of determining the amount of a substrate available to microbial metabolism for calculations of turnover rates of metabolites in sediments. These workers demonstrated that alanine exchanged between a free amino acid pool and an adsorbed or bound pool that appeared to be relatively unavailable to microbial utilization. Such binding may be to clay particles or by complexation with metals. From estimates of the rates of adsorption and desorption of [^{14}C]alanine during their experiments, Christensen and Blackburn (1980) concluded that the adsorbed pool of alanine was 2500 times the free pool. Of the initial rate of alanine removal from the free pool, 27% was by adsorption ($k = 0.06/\text{min}$) and 73% by biological removal ($k = 0.16/\text{min}$). Balba and Nedwell (1982) have similarly observed in cores of autoclaved, biologically inert sediment equilibration of propionate and butyrate, but not acetate, between a free, extractable pool and a bound, nonextractable pool. Adsorption accounted for 63% of the initial rate of removal of [^{14}C]propionate and 78% of removal of [^{14}C]butyrate in unautoclaved cores, although the initial rapid exchange of tracer with the bound pool was complete after 15–20 min and the apparent rate of biological removal could then be estimated. Ansbaek and Blackburn (1980) found, in contrast, that 3% of injected [^{14}C]acetate was adsorbed by Limfjord sediment.

Such exchange makes estimates of turnover rates of metabolites in sediment difficult, since the sizes of the substrate pools being used by the sedimentary microorganisms are often not known and it is uncertain which of the pools are being estimated by the analytical technique used. For example, Christensen and Blackburn (1980) calculated that their estimated rate of alanine mineralization would account for more than the total rate of NH_4^+ release in the Limfjord sediment, although alanine was only a small proportion (7%) of the total amino acids present. They suggested that the discrepancy was because much of the dissolved alanine (800 nM) was unavailable and that a more realistic available pool size would be about 10 nM. Again, the relative availability of each pool to microbial utilization is usually not known, and in short-term experiments, injected radiotracer may turn over rapidly in labile available pools before it is equilibrated with more slowly metabolized, possibly larger, pools of adsorbed substrate. While this is not an argument for not using radiotracers, caution must obviously be applied in interpreting estimates of turnover constants obtained by these methods.

6.4. Acetogenic Bacteria

6.4.1. Proton Reduction and Interspecies H_2 Transfer

Intermediate fermentation products, derived from the activity of hydrolyzing bacteria, may be further metabolized by acetogenic bacteria to yield predominantly acetate together with H_2 and CO_2. An important group of microorganisms at this stage is the "proton-reducing" bacteria (Thauer et al., 1977) involved in interspecies H_2 transfer. This process was first described by Bryant et al. (1967) with "Methanobacillus omelianskii," which was shown to be a consortium of two syntrophic bacteria. Fermentative bacteria normally reoxidize their reduced coenzymes by transfer of electrons to an organic terminal electron acceptor such as pyruvate. In the presence of extremely low concentrations ($< 10^{-4}$ atmosphere) of H_2, however, some bacteria have the ability to reoxidize their reduced coenzymes by the direct release of H_2 (proton reduction) catalyzed by hydrogenase enzyme:

$$NADH_2 \rightarrow NAD + H_2$$

This reaction is feasible only if the environmental H_2 concentration is low enough to permit proton reduction (Wolin, 1976; McInerney and Bryant, 1981), and it is maintained at this low concentration by the H_2-scavenging activity of H_2-metabolizing bacteria such as methanogenic bacteria (MB) or some sulfate-reducing bacteria (SRB). Measurements in anaerobic sediments have shown that H_2 may be present only at the extremely low concentrations that would permit proton reduction and is rapidly turned over (Strayer and Tiedje, 1978; Winfrey et al., 1977; Winfrey and Zeikus, 1979; Jones et al., 1982).

The proton-reducing partner in such syntrophic associations gains a small energetic increment by H_2 release because cellular intermediates such as pyruvate, which would otherwise be reduced in order to reoxidize reduced coenzymes, can be further oxidized to acetate. For example, it was shown in laboratory experiments that Ruminococcus albus cultured alone gave an ATP yield of 3.3 ATP/mole glucose, and ethanol, acetate, CO_2, and H_2 were metabolic products. When cocultured with an H_2-scavenging partner such as Vibrio succinogenes, the ATP yield to R. albus increased to 4 ATP/mole glucose, and the end products contained no ethanol, but only acetate (Iannotti et al., 1973; Tewes and Thauer, 1980). Thus, interspecies H_2 transfer results in a net increase in the energy yield to the proton-reducing bacterium, and there is a decrease in the amount of its reduced fermentation products, such as ethanol or succinate, with a stoichiometrically equal increase in the amount of acetate. A variety of bacterial consortia exhibiting interspecies H_2 transfer have now been described and throw light on the possible reactions that may occur during carbon flow in anaerobic sedimentary communities. It has been demonstrated that a variety of otherwise fermentative bacteria, such as *R. albus,* are able to switch from fermentative to proton-reducing metabolism in the presence of an H_2-scavenging bacterium and low H_2 concentrations (Iannotti *et al.,* 1973; Scheifinger *et al.,* 1975; Latham and Wolin, 1977). Such bacteria are therefore "facultative proton-reducers," and such an ability may be of competitive advantage in an energy-limited environment because of the slightly greater energy yield with proton reduction. However, the cost of metabolic flexibility may be a selective disadvantage if it is not being selected for (Slater and Godwin, 1980), and in environments that are less energy-limited, obligately fermentative bacteria may be less disadvantaged. Moreover, it would appear that many sediments contain only limited amounts of available carbon (energy), and proton reduction is possibly important in such systems.

Other "obligate proton-reducing" bacteria are also known that can metabolize *only* in the presence of an H_2-scavenging partner. The first one recorded was the "S organism" of *"Methanobacillus omelianskii,"* but a number of others have since been described. *Syntrophobacter wolinii* (Boone and Bryant, 1980) was isolated from an anaerobic sewage digester and oxidizes propionate to acetate $+ H_2$ only if cocultured with an H_2-sink organism such as an MB or SRB. *Syntrophomonas wolfei* (McInerney *et al.,* 1981), a bacterium that metabolizes fatty acids of chain lengths up to C_8 by β-oxidation when cocultured with an H_2-sink organism (McInerney *et al.,* 1979), is another. C_{even} fatty acids, such as butyrate, caproate, and caprylate, are oxidized to acetate $+ H_2$ by this bacterium, whereas C_{odd} fatty acids, such as valerate and heptanoate, are oxidized to acetate $+$ propionate $+ H_2$.

It has also been demonstrated (see Evans, 1977) that in the absence of light or nitrate, aromatic compounds such as benzoate can be metabolized anaerobically only if an H_2-sink bacterium is present to remove the metabolic products of ring fission, including fatty acids and H_2. The H_2-sink bacterium can be either an MB or an SRB (Balba and Evans, 1977, 1980). However, it would appear that the degradation of the aromatic rings of lignin does not occur under anaerobic conditions (Hackett *et al.,* 1977; Maccubbin and Hodson, 1980; Zeikus, 1981).

6.4.2. Carbon Flow through Fatty Acid Pools

A wide assortment of fatty acids are found in anaerobic sediments, including long-chain $(>C_{10})$ fatty acids that are derived from structural components of both prokaryotic and eukaryotic cells. However, the most important in relation to anaerobic carbon flow are those short-chain ($<C_6$) fatty acids that are products of bacterial fermentation reactions. These include both the volatile fatty acids, such as acetate, propionate, and butyrate, and the nonvolatile fatty acids, such as succinate and lactate. Available data suggest that in all cases, acetate is the most abundant fatty acid in anaerobic sediments, as it is in other anaerobic environments such as sewage digesters. Reported concentrations of acetate in sediment pore water range from as low as 0.1–6 μM (Winfrey and Zeikus, 1979; Ansbaek and Blackburn, 1980; Balba and Nedwell, 1982) to several hundred micromolar in sediments with high input of organic carbon (Miller *et al.*, 1979; Molongoski and Klug, 1980b; Lovely and Klug, 1982). Propionate and butyrate tend to be the next most abundant fatty acids, whereas there are relatively much lower concentrations of the longer-chain acids, such as valerate and caproate. Cappenberg (1974b) detected lactate in the sediment of Lake Vechten, Holland, where it was metabolized to acetate by sedimentary bacteria (Cappenberg and Prins, 1974; Cappenberg and Jongejan, 1978). Lactate has also been reported in other sediments, generally where there has been a high organic carbon input to the sediment (Miller *et al.*, 1979; Lovely and Klug, 1982). Both lactate and succinate are fermentation products that can be further oxidized to propionate and acetate and are likely to accumulate only where organic carbon is in relatively abundant supply and where proton-reducing bacteria are relatively unimportant.

The metabolism of the predominant short-chain fatty acids by sedimentary bacteria has been investigated using ^{14}C radiotracer techniques. Ansbaek and Blackburn (1980) measured acetate turnover in marine Limfjord sediment and found that the turnover rate constant at depths of 4–6 cm did not vary greatly on a seasonal basis, but was consistent at about 2.1 (± 0.6)/hr. However, variations in the size of the acetate pool within the sediment resulted in seasonal changes in the rates of acetate metabolism. In a saltmarsh pan sediment (Balba and Nedwell, 1982), acetate concentrations diminished with depth over the top 0–22 cm, and the rate of acetate turnover at 6.5°C also diminished from 7.8 nmoles acetate/ml sediment per hr at 0–4 cm to 0 nmole acetate/ml sediment per hr at depths of 18–22 cm. This is consistent with the observation (see Fig. 3) that microbially available carbon was present only at low concentrations below the top 2 cm of sediment.

Examination of acetate, propionate, and butyrate metabolism over the microbiologically most active 0–4 cm horizon of sediment showed that at 6.5°C, the rates were 4.8, 0.3, and 0.19 nmole/ml sediment per hr, respectively. Taking into account the different molar carbon content of each acid, the rates of carbon flow through the sedimentary pool of each acid were in the proportion 12.6:1.2:1.0. Even assuming that all the propionate and butyrate was oxidized to acetate, only 14% of the acetate could have been derived from propionate + butyrate (Balba and Nedwell, 1982).

In sediment from Wintergreen Lake, the turnovers of ^{14}C-labeled acetate, propionate, and lactate were also greatest in the top 0–2 cm horizon (Lovely and Klug, 1982). In the presence of chloroform and molybdate, to inhibit both MB and SRB, which utilize these fatty acid intermediates, 82% of the total fatty acids that accumulated was acetate, 13% propionate, 2% butyrates, and 3% valerates. The relative rates of turnover indicated that as in the United Kingdom saltmarsh sediment, the majority of acetate was derived directly from the initial hydrolysis and metabolism of detrital components, rather than via intermediary fermentation products such as propionate or lactate. It has been shown with laboratory cultures that a variety of substrates, including cellulose (Latham and Wolin, 1977), benzoate (Ferry and Wolfe, 1976), glucose (Iannotti *et al.*, 1973), and fructose (Winter and Wolfe, 1980), may be anaerobically metabolized directly to acetate in the presence of an H_2-scavenging syntrophic bacterium without the intermediate accumulation of reduced fermentation products. These data therefore imply that interspecies H_2 transfer has a major influence in directing the pathway of flow of organic carbon away from intermediate fermentation products, such as propionate, butyrate, and lactate, toward acetate $+ H_2$ during its oxidation in these anaerobic sediments.

6.4.3. Role of Proton-Reducing Bacteria and Sulfate-Reducing Bacteria in Metabolism of Fatty Acids

Until recently, SRB were known to metabolize only a restricted range of short-chain organic molecules, including lactate, pyruvate, ethanol, and glycerol (Postgate, 1979). Widdel (1980) has recently isolated and described a variety of new genera and species of SRB able to metabolize a wide range of fatty acids up to chain lengths of C_{18} and also a variety of aromatic compounds. Many of these new types are also able to directly metabolize propionate and butyrate as electron donors. Hydrogen-scavenging bacteria, either MB or SRB, can also contribute to the degradation of fatty acids, such as propionate and butyrate, by proton-reducing syntrophic fatty-acid-metabolizing bacteria (McInerney *et al.*, 1979, 1981; Boone and Bryant, 1980). In freshwater sediments, SRB may be sulfate-limited and inactive, but in high-sulfate marine sediments, the oxidation of fatty acids could be by either of the mechanisms discussed above. It is possible to differentiate between the two alternatives if H_2 gas is added to inhibit proton reduction by the presence of a high environmental H_2 concentration. Lovely and Klug (1982) used this technique to demonstrate that in the anaerobic sediment from Wintergreen Lake, an atmosphere of H_2 totally inhibited the turnover of propionate, butyrate, isobutyrate, valerate, and isovalerate. This, in the absence of sulfate reduction, is consistent with fatty acid oxidation being by proton-reducing bacteria interacting with H_2-scavenging MB. In similar experiments with slurries of saltmarsh sediment, it has been shown (Banat and Nedwell, 1983) that an H_2 atmosphere did not inhibit propionate and butyrate turnover, although they were completely inhibited under an N_2 atmosphere by 20 mM molybdate, an apparently specific inhibitor of SRB (Peck, 1959; Oremland and Taylor, 1978). This suggests that at least in the high-sulfate sediment from the Colne Point saltmarsh, propionate and butyrate are oxidized directly to acetate by SRB the metabolism of propionate and butyrate of which is not inhibited by H_2, making the mechanism similar to those newly described by Widdel (1980). It will be interesting to see whether future work confirms this as a consistent mechanistic difference for short-chain fatty acid oxidations in low-sulfate and high-sulfate environments.

6.5. Terminal Steps in Anaerobic Organic Carbon Mineralization

6.5.1. Competition for H_2

It was previously thought that in the absence of nitrate, the only possible inorganic product of anaerobic organic carbon mineralization was CH_4, apart from CO_2 formed during fermentations. At the redox potentials found in most anaerobic sediments, nitrate is not present. The two most abundant electron acceptors in anaerobic marine sediments are HCO_3^- and SO_4^{2-}, whereas in anaerobic freshwater sediments, HCO_3^- alone may be present in appreciable quantities. Both the major end products of anaerobic carbon flow (i.e., H_2 and acetate) are known to be utilized by MB to form CH_4 (Wolfe, 1971), whereas only H_2 was known to be metabolized by SRB (see Postgate, 1979).

Studies of the distribution of CH_4 with depth in a variety of sediments had suggested a fundamental difference between marine and freshwater sediments (Reeburgh and Heggie, 1977). Methane was often not detectable within those surface layers of marine sediments where sulfate reduction was most active, whereas in freshwater sediments CH_4 was present throughout the sediment profile, up to the sediment–water interface. Reeburgh and Heggie (1977) suggested that the situation in marine sediments might be the result either of oxidation of CH_4 by the SRB or of the SRB outcompeting the MB for some common environmental resource. In a study of the brackish Lake Vechten, Holland, Cappenberg (1974a) showed that there was an apparent stratification within the sediment, with maximum counts of SRB being found at shallower depths than the maximum counts of MB. In a more recent examination of the distributions of SRB and MB within a number of marine and estuarine sediments (Hines and Buck, 1982), it was found that although in general the distributions of the two groups of bacteria tended to be mutually exclusive, this was not absolute and overlap or occurrence within the same strata of sediment did occur to some extent. Cappenberg (1974a,b) attributed the apparent separation to the toxic effect on the MB of sulfide produced by the SRB. Although sulfide may undoubtedly be toxic to the MB under some circumstances (Cappenberg, 1975; Wellinger and Wuhrmann, 1977), there is little evidence that it is toxic at the concentrations measured within most sediments (Abram and Nedwell, 1978b; Mountfort et al., 1980; Senior et al., 1982).

Cappenberg (1974a) showed that addition of SO_4^{2-} to slurry of Lake Vechten sediment inhibited methanogenesis, an observation repeated in a variety of other low-sulfate sediments (Winfrey and Zeikus, 1977; Zaiss, 1981; Jones et al., 1982). Winfrey and Zeikus (1977) proposed that when sulfate was made available in the normally sulfate-depleted sediment, the SRB were then able to outcompete the MB for common environmental resources. In the absence of sulfate, the SRB were inhibited and methanogenesis occurred. Winfrey and Zeikus (1977) suggested that H_2 or acetate or both were the likely precursors of CH_4 for which there was competition, and they demonstrated that addition of H_2 or acetate to sediment slurry reversed the inhibition of methanogenesis caused by sulfate. These observations supported the idea of competition between MB and SRB for H_2 or acetate or both. Thereafter, Abram and Nedwell (1978a) demonstrated with laboratory cultures that the addition of H_2-utilizing SRB to a syntrophic culture that was actively producing CH_4 inhibited further CH_4 formation, the reason being that the SRB were able to outcompete the MB for H_2 transferred from the proton-reducing syntrophic bacterium. Kristjansson et al. (1982) have found that hydrogenase from Desulfovibrio vulgaris has a lower K_s (1 μM) for H_2 than does hydrogenase from Methanobrevibacterium arborophilus (6 μM), and if these two species are characteristic of SRB and MB, the difference in affinity for H_2 explains why MB are outcompeted by SRB for this resource. In sediment slurries to which excess H_2 was added, both sulfate reduction and methanogenesis were simultaneously stimulated (Abram and Nedwell, 1978b), demonstrating that they were not inherently antagonistic processes, but that the apparent ecological separation resulted from competition.

The importance of SRB as the main H_2-scavenging bacteria in high-sulfate sediments has been demonstrated in a number of studies (Mountfort et al., 1980; Nedwell and Banat, 1981). Inhibition of methanogenesis by inhibitors such as chloroform had no effect on H_2 uptake by slurries of saltmarsh sediment, whereas inhibition of SRB by 20 mM molybdate resulted in H_2 uptake also being totally inhibited (Nedwell and Banat, 1981). In freshwater low-sulfate sediment, though, MB play the equivalent role and are the major H_2 scavengers (Strayer and Tiedje, 1978). Although the SRB may be the most important H_2 scavengers in high-sulfate sediment, however, H_2 may not necessarily be the most important electron donor for sulfate reduction. Sørensen et al. (1981) showed that in a slurry of sediment from a Danish coastal lagoon, H_2 oxidation by SRB accounted for only about 5% of the total amount of sulfate reduced. Although H_2 may be only a minor electron donor for sulfate

[U-^{14}C] acetate in slurries of sediment from the Colne Point saltmarsh, Essex, United Kingdom, was completely inhibited by the addition of 20 mM molybdate, implying that the oxidation of acetate was the result of the activity of the SRB. It is perhaps dangerous to place too much emphasis solely on data obtained with a supposedly specific inhibitor in a mixed microbial community where nonspecific side effects on other parts of the microbial community may also occur. In low-sulfate Wintergreen Lake sediment, for example, sulfate reduction was inhibited by only 0.2 mM molybdate, and concentrations greater than 20 mM inhibited methanogenesis (Smith and Klug, 1981). In other sulfate-depleted sediments, however, 20 mM molybdate had no effect on acetate turnover, which continued unchanged (Banat *et al.*, 1981; Sørensen *et al.*, 1981). Moreover, if 5 mM fluoroacetate, an inhibitor of acetate metabolism, was added to saltmarsh sediment slurry, the sulfate-reduction rate was inhibited by about 60% (Banat *et al.*, 1981), suggesting that acetate oxidation supported about 60% of the total sulfate reduction within the sediment. Sørensen *et al.* (1981) calculated a similar proportion (40–50%) stoichiometrically from the amount of acetate that accumulated in sediment slurry in the presence of added molybdate. They similarly calculated that propionate oxidation accounted for 12%, butyrate + isobutyrate oxidation for 10%, and oxidation of H_2 for only 5% of the total sulfate reduction. In further experiments, Banat *et al.* (1981) showed that under an H_2 atmosphere, the rate of sulfate reduction in sediment slurry was stimulated. Although the addition of fluoroacetate inhibited sulfate reduction, the proportionate inhibition was much smaller than under an N_2 atmosphere. These data were interpreted as showing the presence of two distinct functional groups of SRB. One group are H_2 oxidizers, stimulated by an H_2 atmosphere and not inhibited by fluoroacetate; a second group are acetate oxidizers, with metabolic capabilities similar to those of *Desulfobacter postgateii* (Laanbroek and Pfennig, 1981), which do not metabolize H_2 and are inhibited by fluoroacetate. Both groups of SRB are inhibited by molybdate. Thus, each of the main precursors of CH_4, namely, H_2 and acetate, is competed for by a group of SRB that are able, in the presence of sulfate, to outcompete the equivalent MB for each resource (Fig. 6). It has been reported (Schönheit *et al.*, 1982) that *Desulfobacter postgateii* has a lower K_s for acetate (0.2 mM) than does *Methanosarcina barkeri* (3 mM), and if these organisms are typical of their respective groups, the difference in affinity for acetate provides a mechanism to explain the result of the competition for acetate in high-sulfate sediments.

It would seem that the apparent absence of CH_4 from the surface layers of some marine sediments and the apparent vertical separation of sulfate reduction and methanogenesis are consequences of the competition for common environmental resources in which the SRB outcompete the MB. In the absence of sulfate in freshwater sediments MB predominate, as demonstrated by the much greater proportion of organic carbon mineralization accounted for

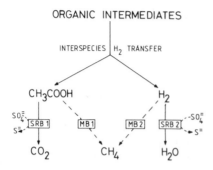

Figure 6. Competition for H_2 and acetate between methanogenic bacteria (MB) and sulfate-reducing bacteria (SRB). Broken lines indicate that in the presence of high sulfate concentrations, MB are outcompeted for resource by SRB. (MB 1) Acetate-utilizing MB; (MB 2) H_2 + CO_2-utilizing MB; (SRB 1) acetate-utilizing SRB; (SRB 2) H_2-utilizing SRB.

by CH_4 formation (see Table II). However, recent evidence (Reeburgh, 1980; Iversen and Blackburn, 1981) indicates that anaerobic oxidation of CH_4 to CO_2 is possible in the sulfate-reduction zone and may also contribute to depletion of CH_4 in the surface layers of marine sediments.

The vertical separation between sulfate reduction and methanogenesis may not be completely exclusive, however, even in high-sulfate marine or intertidal sediments. Hines and Buck (1982) isolated MB from the same strata as SRB in some marine sediments, and Senior *et al.* (1982) have demonstrated that methanogenesis from HCO_3^- proceeds even in the same strata of saltmarsh sediment in which sulfate reduction is most rapid, but at rates that are three orders of magnitude slower than the rates of sulfate reduction. The continuation of methanogenesis, albeit at extremely low rates, was attributed to the MB persisting by obtaining a very small proportion of available H_2 even in the presence of active sulfate reduction. Even though the SRB have a greater affinity for H_2 and acquire the major part of it, the MB will still acquire a small proportion, and in the absence of the physical removal that would occur by washout in a chemostat, they can persist at very low rates of activity within a sediment.

6.5.3. Methane Precursors Other Than H_2 + CO_2 or Acetate

In an examination of CH_4 formation in saltmarsh sediment, Senior *et al.* (1982) concluded that the amount of CH_4 emitted from the sediment surface could not be accounted for by the amount measured as being formed from H_2 + CO_2 within the sediment profile. (Acetate was not a CH_4 precursor in this high-sulfate sediment.) They suggested that precursors other than acetate or H_2 + CO_2, the conversion of which to CH_4 was not measured, might account for the discrepancy. *Methanosarcina barkeri* and *Methanococcus mazei,* for example, are known to form CH_4 from other precursors, including methylated amines (Hippe *et al.,* 1979), methanol, formaldehyde, and formate (Wolfe, 1971; Wolfe and Higgins, 1979). Methane is also formed in sediment from the terminal methyl group of methionine (Zinder and Brock, 1978).

With anaerobic sediment from Big Soda Lake, Nevada, Oremland *et al.* (1982a) found that addition of both methanol and trimethylamine, and to a lesser extent addition of methionine, stimulated CH_4 formation. There was only slight, or no, stimulation by added H_2, acetate, or formate, however. Methanol is known to be a major product of anaerobic breakdown of pectins (Schink and Zeikus, 1980, 1982), whereas methionine is derived from protein hydrolysis. Methylated amines are formed during decomposition of choline, creatine, beatine, and other aminated organic molecules. In similar work with a United States saltmarsh sediment, methanogenesis was again stimulated by addition of methanol or trimethylamine, but these did not apparently stimulate sulfate reduction (Oremland *et al.*, 1982b). Oremland and co-workers suggested that the metabolism by MB of CH_4 precursors, which were not metabolized or competed for by SRB, permitted the continuation of methanogenesis even within high-sulfate environments. Banat *et al.* (1983) also found that when sulfate reduction in a slurry of sediment from a United Kingdom saltmarsh was inhibited by molybdate, methanogenesis was increased. Although the $H_2 + CO_2$ pathway to CH_4 was stimulated to some extent, this source did not account for all the additional CH_4 formed. The turnover of [14C]methanol, [14C]formaldehyde, and [methyl-14C]methionine to $^{14}CH_4$ was stimulated in the presence of molybdate, but no methanogenesis from these precursors was measured in the absence of molybdate, that is, when sulfate reduction was active. [As in the Big Soda Lake study (Oremland *et al.*, 1982a), formate was not converted to CH_4 at all.] These data imply that SRB were also involved in the turnover of the pools of methanol, formaldehyde, and methionine within the United Kingdom saltmarsh sediment, in contrast to the conclusion of Oremland *et al.* (1982b) that methanol was not being used by SRB in the United States saltmarsh sediment. It has been reported previously (Postgate, 1979) that methanol can be used as an electron donor by SRB, but further information is required to resolve this apparent contradiction between two different sedimentary sites.

Again, the observation (Banat *et al.*, 1983) that methionine was normally oxidized to CO_2 in saltmarsh sediment, but in the presence of 20 mM molybdate was converted to CH_4, is interesting because it implies that methionine was normally metabolized by SRB. Senez and Leroux-Gitteron [(1954) (cited by Postgate, 1979)] showed that a marine strain of *Desulfovibrio desulfuricans* oxidatively degraded the other thioamino acid cysteine in the presence of sulfate to acetate, CO_2, NH_3, and H_2S. In the absence of sulfate, cysteine was deaminated, and the subsequent products of degradation were similar to those of pyruvate. Smith and Klug (1981) also found that addition of 0.2 mM molybdate to sediment from the profundal region of Wintergreen Lake decreased mineralization of an amino acid mixture by 85%, and this they attributed to the inhibition of SRB activity.

Ingvorsen *et al.* (1981) investigated the response of sulfate reduction in sediment from Lake Mendota to additions of various electron donors. The degree of stimulation by each addition was in the order $H_2 > n$-butanol $> n$-propanol $>$ ethanol $>$ glucose $= n$-butyrate $> n$-propionate. This indicates that alcohols are at least potentially important intermediates in anaerobic carbon flow, apart from fatty acids. Although methanol and ethanol have been detected in a number of sediments (Oremland et al., 1982b; Banat, unpublished data), there is no information available regarding either sizes of alcohol pools within sediments or turnover rates.

Other intermediates commonly formed during anaerobic organic carbon degradation, particularly in high-sulfate environments, are organic sulfur compounds such as thiols (Francis et al., 1975; Zinder and Brock, 1978). Zinder and Brock (1978) demonstrated that methane thiol (CH_3SH), an analogue of methanol, accumulated in Lake Mendota sediment in the presence of chloroform. Methionine labeled with ^{35}S gave rise to other volatile organic sulfur compounds such as dimethylsulfide and dimethyldisulfide. The same workers (unpublished data, cited in Zinder and Brock, 1978) showed that [^{14}C]methane thiol was metabolized to $^{14}CH_4$ by sedimentary bacteria. The quantitative contribution of such intermediates to CH_4 formation is at present unknown. The data emphasize, however, that much more information is required on the pool sizes and turnover rates of a wide variety of intermediary products of organic carbon oxidation in anaerobic sediments before the pathways, and the environmental factors regulating the pathways, for organic carbon flow in these environments are fully understood.

ACKNOWLEDGMENTS. I would like to acknowledge the collaboration of many colleagues in the work on Colne Point saltmarsh sediments that has been discussed herein, and the Natural Environment Research Council of the United Kingdom for financial support over a number of years.

References

Abd. Aziz, S. A., and Nedwell, D. B., 1979, Microbial nitrogen transformations in the saltmarsh environment, in: Ecological Processes in the Coastal Environment (R. L. Jefferies and A. J. Davy, eds.), pp. 385–398, Blackwell, Oxford.

Abram, J. W., and Nedwell, D. B., 1978a, Inhibition of methanogenesis by sulphate reducing bacteria competing for transferred hydrogen, Arch. Microbiol. 117:89–92.

Abram, J. W., and Nedwell, D. B., 1978b, Hydrogen as a substrate for methanogenesis and sulphate reduction in anaerobic saltmarsh sediment, Arch. Microbiol. 117:93–97.

Aller, R. C., and Yingst, J. Y., 1980, Relationships between microbial distributions and the anaerobic decomposition of organic matter in surface sediments of Long Island Sound, USA, Mar. Biol. 56:29–42.

Aller, R. C., Benninger, L. K., and Cochran, J. K., 1980, Tracking particle-associated processes in nearshore environments by use of $^{234}Th/^{238}U$ disequilibrium, Earth Planet. Sci. Lett. 47:161–175.

Ansbaek, J., and Blackburn, T. H., 1980, A method for the analysis of acetate turnover in a coastal marine sediment, *Microb. Ecol.* **5:**253–264.

Ansell, A. D., 1974, Sedimentation of organic detritus in Lochs Etive and Creran, Argyll, Scotland, *Mar. Biol.* **27:**263–273.

Balba, M. T., and Evans, W. C., 1977, The methanogenic fermentation of aromatic substrates, *Biochem. Soc. Trans.* **5:**302–304.

Balba, M. T., and Evans, W. C., 1980, The anaerobic dissimilation of benzoate by *Pseudomonas aeruginosa* coupled with *Desulfovibrio vulgaris,* with sulphate as terminal electron acceptor, *Biochem. Soc. Trans.* **8:**624–625.

Balba, M. T., and Nedwell, D. B., 1982, Microbial metabolism of acetate, propionate and butyrate in anoxic sediment from Colne Point saltmarsh, Essex, U.K., *J. Gen. Microbiol.* **128:**1415–1422.

Banat, I. M., and Nedwell, D. B., 1983, Mechanisms of turnover of C_2-C_4 fatty acids in high-sulphate and low-sulphate anaerobic sediments, *FEMS Microbiol. Lett.* **17:**107–110.

Banat, I. M., Lindström, E. B., Nedwell, D. B., and Balba, M. T., 1981, Evidence for coexistence of two distinct functional groups of sulfate-reducing bacteria in salt marsh sediment, *Appl. Environ. Microbiol.* **42:**985–992.

Banat, I. M., Nedwell, D. B., and Balba, M. T., 1983, Stimulation of methanogenesis by slurries of saltmarsh sediment by addition of molybdate to inhibit sulphate-reducing bacteria, *J. Gen. Microbiol.* **129:**123–129.

Barber, R. T., 1968, Dissolved organic carbon from deep waters resists microbial oxidation, *Nature (London)* **220:**274–275.

Benoit, G. J., Turekian, K. K., and Benninger, L. K., 1979, Radiocarbon dating of a core from Long Island Sound, *Estuarine Coastal Mar. Sci.* **9:**171–180.

Berner, R. A., 1978, Sulfate reduction and the rate of deposition of marine sediments, *Earth Planet. Sci. Lett.* **37:**492–498.

Berner, R. A., 1980a, *Early Diagenesis—A Theoretical Approach,* Princeton University Press, Princeton, New Jersey.

Berner, R. A., 1980b, A rate model for organic matter decomposition during bacterial sulphate reduction in marine sediments, in: *Biogéochemie de la Matière Organique a l'Interface Eau–Sediment Marin* (R. Daumas, ed.), pp. 35–45, CNRS, Paris.

Billen, G., 1982, Modelling the processes of organic matter degradation and nutrients recycling in sedimentary systems, in: *Sediment Microbiology* (D. B. Nedwell and C. M. Brown, eds.), pp. 15–52, Academic Press, London.

Billen, G., and Verbeustel, S., 1980, Distribution of microbial metabolisms in natural environments displaying gradients of oxidation–reduction reactions, in: *Biogéochemie de la Matière Organique a l'Interface Eau–Sediment Marin* (R. Daumas, ed.), pp. 291–300, CNRS, Paris.

Bloesch, J., Stadelmann, P., and Buhrer, H., 1977, Primary production, mineralization, and sedimentation in the euphotic zone of two Swiss lakes, *Limnol. Oceanogr.* **22:**511–526.

Boone, D. R., and Bryant, M. P., 1980, Propionate-degrading bacterium, *Syntrophobacter wolinii* sp. nov. gen. nov., from methanogenic ecosystems, *Appl. Environ. Microbiol.* **40:**626–632.

Bouldin, D. R., 1968, Models for describing the diffusion of oxygen and other mobile constituents across the mud–water interface, *J. Ecol.* **56:**77–87.

Bryant, M. P., Wolin, E. A., Wolin, M. J., and Wolfe, R. S., 1967, *Methanobacillus omelianskii,* a symbiotic association of two species of bacteria, *Arch. Microbiol.* **59:**20–31.

Burns, R. G., 1978, *Soil Enzymes,* Academic Press, London.

Cappenberg, T. E., 1974a, Interrelations between sulphate-reducing and methane-producing bacteria in bottom deposits of a freshwater lake. I. Field observations, *Antonie van Leeuwenhoek J. Microbiol. Serol.* **40:**285–295.

Cappenberg, T. E., 1974b, Interrelations between sulphate-reducing and methane-producing bacteria in bottom deposits of a freshwater lake. II. Inhibition experiments, *Antonie van Leeuwenhoek J. Microbiol. Serol.* **40:**297–306.

Cappenberg, T. E., 1975, A study of mixed continuous cultures of sulphate-reducing and methane-producing bacteria, *Microb. Ecol.* **2:**60–72.

Cappenberg, T. E., and Jongejan, E., 1978, Microenvironments for sulfate reduction and methane production in freshwater sediments, in: *Environmental Biogeochemistry and Geomicrobiology,* Vol. 1 (W. E. Krumbein, ed.), pp. 129–138, Ann Arbor Science Publishers, Ann Arbor, Michigan.

Cappenberg, T. E., and Prins, R. A., 1974, Interrelations between sulphate-reducing and methane-producing bacteria in bottom deposits of a freshwater lake. III. Experiments with ^{14}C-labelled substrates, *Antonie van Leeuwenhoek J. Microbiol. Serol.* **40:**457–469.

Christensen, D., and Blackburn, T. H., 1980, Turnover of tracer (^{14}C, ^{3}H labelled) alanine in inshore marine sediments, *Mar. Biol.* **58:**97–103.

Christian, R. R., and Wiebe, W. J., 1978, Anaerobic microbial community metabolism in *Spartina alterniflora* soils, *Limnol. Oceanogr.* **23:**328–336.

Davies, J. M., 1975, Energy flow through the benthos in a Scottish sea loch, *Mar. Biol.* **31:**353–362.

Degens, E. T., and Mopper, K., 1976, Factors controlling the distribution and early diagenesis of organic material in marine sediments, in: *Chemical Oceanography* (J. P. Riley and R. Chester, eds.), pp. 59–113, Academic Press, London.

De la Cruz, A. A., 1975, Proximate nutritive value changes during decomposition of saltmarsh plants, *Hydrobiologia* **47:**475–480.

Evans, W. C., 1977, Biochemistry of the bacterial catabolism of aromatic compounds in anaerobic environments, *Nature (London)* **270:**17–22.

Fallon, R. D., and Pfaender, F. K., 1976, Carbon metabolism in model microbial systems from a temperate salt marsh, *Appl. Environ. Microbiol.* **31:**959–968.

Fallon, R. D., Harrits, S., Hanson, R. S., and Brock, T. D., 1980, The role of methane in internal carbon cycling in Lake Mendota during summer stratification, *Limnol. Oceanogr.* **25:**357–360.

Fenchel, T., and Blackburn, T. H., 1979, *Bacteria and Mineral Cycling,* Academic Press, London.

Fenchel, T., and Harrison, P., 1976, The significance of bacterial grazing and mineral cycling for the decomposition of particulate detritus, in: *The Role of Terrestrial and Aquatic Organisms in Decomposition Processes* (J. M. Anderson and A. Macfadyen, eds.), pp. 285–299, Blackwell, Oxford.

Fenchel, T., and Jørgensen, B. B., 1977, Detritus food chains of aquatic ecosystems: The role of bacteria, *Adv. Microb. Ecol.* **1:**1–58.

Ferry, J. G., and Wolfe, R. S., 1976, Anaerobic degradation of benzoate to methane by a microbial consortium, *Arch. Microbiol.* **107:**33–40.

Fogg, G. E., 1966, The extracellular products of algae, *Oceanogr. Mar. Biol. Annu. Rev.* **4:**195–212.

Francis, A. J., Duxbury, J. M., and Alexander, M., 1975, Formation of volatile organic products in soils under anaerobiosis. II. Metabolism of amino acids, *Soil Biol. Biochem.* **7:**51–56.

Froelich, P. N., Klinkhammer, G. P., Bender, M. L., Luedtke, N. A., Heath, G. R., Cullen, D., Dauphin, P., Hammond, D., Hartman, B., and Maynard, V., 1979, Early oxidation of organic matter in pelagic sediments of the eastern equatorial Atlantic: Suboxic diagenesis, *Geochim. Cosmochim. Acta* **43:**1075–1090.

Fry, J. C., 1982, Interactions between bacteria and benthic invertebrates, in: *Sediment Microbiology* (D. B. Nedwell and C. M. Brown, eds.), pp. 171–201, Academic Press, London.

Godshalk, G. L., and Wetzel, R. G., 1978a, Decomposition of aquatic angiosperms. I. Dissolved components, *Aquat. Bot.* **5:**281–300.

Godshalk, G. L., and Wetzel, R. G., 1978b, Decomposition of aquatic angiosperms. II. Particulate components, *Aquat. Bot.* **5**:301–327.

Godshalk, G. L., and Wetzel, R. G., 1978c, Decomposition of aquatic angiosperms. III. *Zostera marina* L. and a conceptual model of decomposition, *Aquat. Bot.* **5**:329–354.

Hackett, W. F., Connors, W. J., Kirk, T. K., and Zeikus, J. G., 1977, Microbial decomposition of synthetic ^{14}C-labelled lignins in nature: Lignin biodegradation in a variety of natural materials, *Appl. Environ. Microbiol.* **33**:43–51.

Hall, K. J., Kleiber, P. M., and Jesaki, I., 1972, Heterotrophic uptake of organic solutes by microorganisms in the sediment, *Mem. Ist. Ital. Idrobiol. Suppl.* **29**:441–471.

Hanson, R. B., and Gardner, W. S., 1978, Uptake and metabolism of two amino acids by anaerobic microorganisms in four diverse salt marsh soils, *Mar. Biol.* **46**:101–108.

Hargrave, B. T., 1972, Aerobic decomposition of sediment and detritus as a function of particle surface area and organic content, *Limnol. Oceanogr.* **17**:583–596.

Hines, M. E., and Buck, J. D., 1982, Distribution of methanogenic and sulfate-reducing bacteria in near-shore marine sediments, *Appl. Environ. Microbiol.* **43**:447–453.

Hippe, H., Caspari, D., Fiebig, K., and Gottschalk, G., 1979, Utilization of trimethylamine and other N-methyl compounds for growth and methane formation by *Methanosarcina barkeri, Proc. Natl. Acad. Sci. U.S.A.* **76**:494–498.

Holm-Hansen, O., 1972, The distribution and chemical composition of particulate material in marine and freshwaters, *Mem. Ist. Ital. Idrobiol. Suppl.* **29**:37–51.

Honjo, S., 1978, Sedimentation of materials in the Sargasso Sea at a 5367 m deep station, *J. Mar. Res.* **36**:469–492.

Honjo, S., and Roman, M. R., 1978, Marine copepod faecal pellets: Production, preservation and sedimentation, *J. Mar. Res.* **36**:45–57.

Howeller, R. H., 1972, The oxygen status of lake sediments, *J. Environ. Qual.* **1**:366–371.

Hylleberg, J., and Henriksen, K., 1980, The central role of bioturbation in sediment mineralization and element recycling, *Ophelia Suppl.* **1**:1–16.

Iannotti, E. L., Kafkewitz, C., Wolin, M. J., and Bryant, M. P., 1973, Glucose fermentation products of *Ruminococcus albus* grown in continuous culture with *Vibrio succinogenes:* Changes caused by interspecies transfer of H_2, *J. Bacteriol.* **114**:1231–1240.

Ingvorsen, K., Zeikus, J. G., and Brock, T. D., 1981, Dynamics of bacterial sulfate reduction in a eutrophic lake, *Appl. Environ. Microbiol.* **42**:1029–1036.

Iturriaga, R., 1979, Bacterial activity related to sedimenting particulate matter, *Mar. Biol.* **55**:157–169.

Iversen, N., and Blackburn, T. H., 1981, Seasonal rates of methane oxidation in anoxic marine sediments, *Appl. Environ. Microbiol.* **41**:1295–1300.

Jones, J. G., 1971, Studies on freshwater bacteria: Factors which influence the population and its activity, *J. Ecol.* **59**:593–613.

Jones, J. G., 1976, The microbiology and decomposition of seston in open water and experimental enclosures in a productive lake, *J. Ecol.* **64**:241–278.

Jones, J. G., 1982, Activities of aerobic and anaerobic bacteria in lake sediments and their effect on the water column, in: *Sediment Microbiology* (D. B. Nedwell and C. M. Brown, eds.), pp. 107–145, Academic Press, London.

Jones, J. G., and Simon, B. M., 1980, Decomposition processes in the profundal region of Blelham Tarn and the Lund Lakes, *J. Ecol.* **68**:493–512.

Jones, J. G., and Simon, B. M., 1981, Differences in microbial decomposition processes in profundal and littoral lake sediments, with particular reference to the nitrogen cycle, *J. Gen. Microbiol.* **123**:297–312.

Jones, J. G., Simon, B. M., and Gardner, S., 1982, Factors affecting methanogenesis and associated anaerobic processes in the sediments of a stratified eutrophic lake, *J. Gen. Microbiol.* **128**:1–11.

Jørgensen, B. B., 1978, A comparison of methods for the quantification of bacterial sulfate reduction in coastal marine sediments. II. Calculation from mathematical models, *Geomicrobiol. J.* **1**:29–47.

Jørgensen, B. B., 1980, Mineralization and the bacterial cycling of carbon, nitrogen and sulphur in marine sediments, in: *Contemporary Microbial Ecology* (D. C. Ellwood, J. N. Hedger, M. J. Latham, J. M. Lynch, and J. H. Slater, eds.), pp. 239–252, Academic Press, London.

Jørgensen, B. B., 1982, Mineralization of organic matter in the sea bed—the role of sulphate reduction, *Nature (London)* **296**:643–645.

Khailov, K. M., and Burlakova, Z. P., 1969, Release of dissolved organic matter by marine seaweeds and distribution of their total organic production to inshore communities, *Limnol. Oceanogr.* **14**:521–527.

King, G. M., and Wiebe, W. J., 1980, Tracer analysis of methanogenesis in saltmarsh soil, *Appl. Environ. Microbiol.* **39**:877–881.

Kristjansson, J. K., Schönheit, P., and Thauer, R. K., 1982, Different K_s values for hydrogen of methanogenic bacteria and sulfate reducing bacteria, *Arch. Microbiol.* **131**:278–282.

Laanbroek, H. J., and Pfennig, N., 1981, Oxidation of short chain fatty acids by sulfate-reducing bacteria in freshwater and marine sediments, *Arch. Microbiol.* **128**:330–335.

Lastein, E., 1976, Recent sedimentation and resuspension of organic matter in eutrophic Lake Esrom, Denmark, *Oikos* **27**:44–49.

Latham, M. J., and Wolin, M. J., 1977, Fermentation of cellulose by *Ruminococcus flavefaciens* in the presence and absence of *Methanobacterium ruminantium*, *Appl. Environ. Microbiol.* **34**:297–301.

Lee, J. J., 1980, A conceptual model of marine detrital decomposition and the organisms associated with the process, *Adv. Aquat. Microbiol.* **2**:257–291.

Lovely, D. R., and Klug, M. J., 1982, Intermediary metabolism of organic matter in the sediments of a eutrophic lake, *Appl. Environ. Microbiol.* **43**:552–560.

Maccubbin, A. E., and Hodson, R. E., 1980, Mineralization of detrital lignocelluloses by salt marsh sediment microflora, *Appl. Environ. Microbiol.* **40**:735–740.

Mann, K. H., 1976, Decomposition of marine macrophytes, in: *The Role of Terrestrial and Aquatic Organisms in Decomposition Processes* (J. M. Anderson and A. Macfadyen, eds.), pp. 247–267, Blackwell, Oxford.

McInerney, M. J., and Bryant, M. P., 1981, Basic principles of bioconversions in anaerobic digestion and methanogenesis, in: *Biomass Conversion Processes for Energy and Fuels* (S. S. Sofer and O. R. Zaborsky, eds.), pp. 277–296, Plenum Press, New York.

McInerney, M. J., Bryant, M. P., and Pfennig, N., 1979, Anaerobic bacterium that degrades fatty acids in syntrophic association with methanogens, *Arch. Microbiol.* **122**:129–135.

McInerney, M. J., Bryant, M. P., Hespell, R. B., and Costerton, J. W., 1981, *Syntrophomonas wolfei* gen. nov. sp. nov., an anaerobic, syntrophic, fatty acid-oxidizing bacterium, *Appl. Environ. Microbiol.* **41**:1029–1039.

Mechalas, B. J., 1974, Pathways and environmental requirements for biogenic gas production in the oceans, in: *Natural Gases in Marine Sediments* (I. R. Kaplan, ed.), pp. 12–25, Plenum Press, New York.

Menzel, D. W., 1966, Bubbling of seawater and the production of organic particles: A revaluation, *Deep-Sea Res.* **13**:963–966.

Miller, D., Brown, C. M., Pearson, T. H., and Stanley, S. O., 1979, Some biologically important low molecular weight organic acids in the sediments of Loch Eil, *Mar. Biol.* **50**:375–383.

Molongoski, J. J., and Klug, M. J., 1976, Characterization of anaerobic heterotrophic bacteria isolated from freshwater lake sediments, *Appl. Environ. Microbiol.* **31**:83–90.

Molongoski, J. J., and Klug, M. J., 1980a, Quantification and characterization of sedimenting particulate organic matter in a shallow hypereutrophic lake, *Freshwater Biol.* **10**:497–506.

Molongoski, J. J., and Klug, M. J., 1980b, Anaerobic metabolism of particulate organic matter in the sediments of a hypereutrophic lake, *Freshwater Biol.* **10**:507–518.

Mopper, K., 1980, Carbohydrates in the marine environment: Recent developments, in: *Biogéo-chemie de la Matière Organique a l'Interface Eau–Sediment Marin* (R. Daumas, ed.), pp. 35–45, CNRS, Paris.

Morris, J. G., 1975, The physiology of obligate anaerobiosis, *Adv. Microb. Physiol.* **12**:169–246.

Mountfort, D. O., and Asher, R. A., 1981, Role of sulfate reduction versus methanogenesis in terminal carbon flow in polluted intertidal sediment of Waimea Inlet, Nelson, New Zealand, *Appl. Environ. Microbiol.* **42**:252–258.

Mountfort, D. O., Asher, R. A., Mays, E. L., and Tiedje, J. M., 1980, Carbon and electron flow in mud and sandflat intertidal sediments at Delaware Inlet, Nelson, New Zealand, *Appl. Environ. Microbiol.* **39**:686–694.

Nedwell, D. B., and Banat, I. M., 1981, Hydrogen as an electron donor for sulfate-reducing bacteria in slurries of salt marsh sediment, *Microb. Ecol.* **7**:305–313.

Nelson, D. R., and Zeikus, J. G., 1974, Rapid method for the radioisotopic analysis of gaseous end products of anaerobic metabolism, *Appl. Microbiol.* **28**:258–261.

Odum, E. P., and de la Cruz, A. A., 1967, Particulate organic detritus in a Georgia salt marsh-estuarine system, in: *Estuaries* (G. H. Lauff, ed.), pp. 383–390, American Association for the Advancement of Science, Washington, D.C.

Oppenheimer, C. H., 1960, Bacterial activity in sediments of shallow marine bays, *Geochim. Cosmochim. Acta* **19**:244–260.

Oremland, R. S., and Taylor, B. F., 1978, Sulphate reduction and methanogenesis in marine sediments, *Geochim. Cosmochim. Acta* **42**:209–214.

Oremland, R. S., Marsh, L., and DesMarais, D. J., 1982a, Methanogenesis in Big Soda Lake, Nevada: An alkaline, moderately hypersaline desert lake, *Appl. Environ. Microbiol.* **43**:462–468.

Oremland, R. S., Marsh, L. M., and Polcin, S., 1982b, Methane production and simultaneous sulphate reduction in anoxic salt marsh sediments, *Nature (London)* **296**:143–145.

Otsuki, A., and Hanya, T., 1972a, Production of dissolved organic matter from dead green algal cells. I. Aerobic microbial decomposition, *Limnol. Oceanogr.* **17**:248–264.

Otsuki, A., and Hanya, T., 1972b, Production of dissolved organic matter from dead green algal cells. II. Anaerobic microbial decomposition, *Limnol. Oceanogr.* **17**:258–264.

Parsons, T. R., 1963, Suspended organic matter in seawater, in: *Progress in Oceanography*, Vol. 1 (M. Sears, ed.), pp. 205–239, Pergamon Press, Oxford.

Peck, H. D., 1959, The ATP-dependent reduction of sulphate with hydrogen in extracts of *Desulfovibrio desulfuricans*, *Proc. Natl. Acad. Sci. U.S.A.* **45**:701–708.

Pennington, W., 1974, Seston and sediment formation in five Lake District lakes, *J. Ecol.* **62**:215–251.

Platt, H. M., 1979, Sedimentation and the distribution of organic matter in a sub-Antarctic marine bay, *Estuar. Coastal Mar. Sci.* **9**:51–62.

Postgate, J. R., 1979, *The Sulphate-Reducing Bacteria*, Cambridge University Press, Cambridge.

Qasim, S. Z., and Sankaranarayanan, V. V., 1972, Organic detritus of a tropical estuary, *Mar. Biol.* **15**:193–199.

Redfield, A. C., Ketchum, B. H., and Richards, F. A., 1963, The influence of organisms on the composition of sea-water, in: *The Sea*, Vol. 2 (M. N. Hill, ed.), pp. 26–77, Interscience, New York.

Reeburgh, W. S., 1980, Anaerobic methane oxidation: Rate depth distributions in Skan Bay sediments, *Earth Planet. Sci. Lett.* **47**:345–352.

Reeburgh, W. S., and Heggie, D. T., 1977, Methane consumption reactions and their effect on methane distributions in freshwater and marine environments, *Limnol. Oceanogr.* **22**:1–9.

Revsbech, N. P., Sørensen, J., and Blackburn, T. H., 1980, Distribution of oxygen in marine sediments measured with microelectrodes, *Limnol. Oceanogr.* **25**:403–411.

Rowe, G. T., and Gardner, W. D., 1979, Sedimentation rates in the slope water of the northwest Atlantic Ocean measured directly with sediment traps, *J. Mar. Res.* **37**:581–600.

Saunders, G. W., 1976, Decomposition in freshwater, in: *The Role of Terrestrial and Aquatic Organisms in Decomposition Processes* (J. M. Anderson and A. Macfadyen, eds.), pp. 341–373, Blackwell, Oxford.

Scheifinger, C. C., Lineham, B., and Wolin, M. J., 1975, H_2 production by *Selenomonas ruminantium* in the absence and presence of methanogenic bacteria, *Appl. Microbiol.* **29**:480–483.

Schink, B., and Zeikus, J. G., 1980, Microbial methanol formation: A major end product of pectin metabolism, *Curr. Microbiol.* **4**:387–390.

Schink, B., and Zeikus, J. G., 1982, Microbial ecology of pectin decomposition in anoxic lake sediments, *J. Gen. Microbiol.* **128**:393–404.

Schönheit, P., Kristjansson, J. K., and Thauer, R. K., 1982, Kinetic mechanism for the ability of sulfate reducers to outcompete methanogens for acetate, *Arch. Microbiol.* **132**:285–288.

Seki, H., Skelding, J., and Parsons, T. R., 1968, Observations on the decomposition of a marine sediment, *Limnol. Oceanogr.* **13**:440–448.

Senez, J. C., and Leroux-Gitteron, J., 1954, Preliminary note on the anaerobic degradation of cysteine and cystine by sulphate-reducing bacteria, *Bull. Soc. Chim. Biol.* **36**:553–559.

Senior, E., Lindström, E. B., Banat, I. M., and Nedwell, D. B., 1982, Sulfate reduction and methanogenesis in the sediment of a saltmarsh on the east coast of the United Kingdom, *Appl. Environ. Microbiol.* **43**:987–996.

Slater, J. H., and Godwin, D., 1980, Microbial adaptation and selection, in: *Contemporary Microbial Ecology* (D. C. Ellwood, J. N. Hedger, M. J. Latham, J. M. Lynch, and J. H. Slater, eds.), pp. 137–160, Academic Press, London.

Smith, R. L., and Klug, M. J., 1981, Electron donors utilized by sulfate-reducing bacteria in eutrophic lake sediments, *Appl. Environ. Microbiol.* **42**:116–121.

Sørensen, J., Christensen, D., and Jørgensen, B. B., 1981, Volatile fatty acids and hydrogen as substrates for sulfate-reducing bacteria in anaerobic marine sediments, *Appl. Environ. Microbiol.* **42**:5–11.

Steele, J. H., 1974, *The Structure of Marine Ecosystems,* Harvard University Press, Cambridge, Massachusetts.

Stephens, K., Sheldon, R. W., and Parsons, T. R., 1967, Seasonal variations in the availability of food for benthos in a coastal environment, *Ecology* **48**:852–855.

Strayer, R. F., and Tiedje, J. M., 1978, Kinetic parameters of the conversion of methane precursors to methane in a hypereutrophic lake sediment, *Appl. Environ. Microbiol.* **36**:330–340.

Suess, E., and Muller, P. J., 1980, Productivity, sedimentation rate and sedimentary organic matter in the oceans. II. Elemental fractionation, in: *Biogéochemie de la Matière Organique a l'Interface Eau–Sediment Marin* (R. Daumas, ed.), pp. 17–26, CNRS, Paris.

Taguchi, S., 1982, Sedimentation of newly produced particulate organic matter in a subtropical inlet, Kaneohe Bay, Hawaii, *Estuarine Coastal Shelf Sci.* **14**:533–544.

Tewes, F. J., and Thauer, R. K., 1980, Regulation of ATP-synthesis in glucose-fermenting bacteria involved in interspecies hydrogen transfer, in: *Anaerobes and Anaerobic Infections* (G. Gottschalk, N. Pfennig, and H. Werner, eds.), pp. 97–104, Gustav Fischer Verlag, Stuttgart.

Thauer, R. K., Jungermann, K., and Decker, K., 1977, Energy conservation in chemotrophic anaerobic bacteria, *Bacteriol. Rev.* **41**:100–180.

Toth, D. J., and Lerman, A., 1977, Organic matter reactivity and sedimentation rates in the ocean, *Am. J. Sci.* **277**:465–485.

Turekian, K. K., Benoit, G. J., and Benninger, L. K., 1980, The mean residence time of plankton-derived carbon in a Long Island Sound sediment core: A correction, *Estuarine Coastal Mar. Sci.* **11**:583.

Vigneaux, M., Dumon, J. C., Faugeres, J. C., Grousset, F., Jouanneau, J. M., Latouche, C., Poutiers, J., and Pujol, C., 1980, Matières organiques et sedimentation en milieu marin, in: *Biogéochemie de la Matière Organique a l'Interface Eau–Sediment Marin* (R. Daumas, ed.), pp. 113–128, CNRS, Paris.

Webster, T. J. M., Paranjape, M. A., and Mann, K. H., 1975, Sedimentation of organic matter in St. Margarets Bay, Nova Scotia, *J. Fish. Res. Board Canada* **32:**1399–1407.

Wellinger, A., and Wuhrmann, K., 1977, Influence of sulfide compounds on the metabolism of *Methanobacterium* strain AZ, *Arch. Microbiol.* **115:**13–17.

Wetzel, R. G., Rich, P. R., Miller, M. C., and Allen, H. L., 1972, Metabolism of dissolved and particulate detrital carbon in a temperate hard-water lake, *Mem. Ist. Ital. Idrobiol. Suppl.* **29:**185–243.

Widdel, F., 1980, Anaerober Abbau von Fettsäuren und Benzoesäure durch neu isolierte Arten sulfat-reduzierender Bakterien, Doctoral thesis, University of Göttingen, FDR.

Wiebe, P. H., Boyd, S. H., and Winget, C., 1976, Particulate matter sinking to the deep-sea floor at 2000 m in the Tongue of the Ocean, Bahamas, with a description of a new sedimentation trap, *J. Mar. Res.* **34:**341–354.

Winfrey, M. R., and Zeikus, J. G., 1977, Effect of sulfate on carbon and electron flow during microbial methanogenesis in freshwater sediments, *Appl. Environ. Microbiol.* **33:**275–281.

Winfrey, M. R., and Zeikus, J. G., 1979, Anaerobic metabolism of immediate methane precursors in Lake Mendota, *Appl. Environ. Microbiol.* **37:**244–253.

Winfrey, M. R., Nelson, D. R., Klevickis, S. C., and Zeikus, J. G., 1977, Association of hydrogen metabolism with methanogenesis in Lake Mendota sediments, *Appl. Environ. Microbiol.* **33:**312–318.

Winter, J., and Wolfe, R. S., 1980, Methane formation from fructose by syntrophic association of *Acetobacterium woodii* and different strains of methanogens, *Arch. Microbiol.* **124:**73–79.

Wolfe, R. S., 1971, Microbial formation of methane, *Adv. Microb. Physiol.* **6:**107–146.

Wolfe, R. S., and Higgins, I. J., 1979, Microbial biochemistry of methane—a study in contrasts: Microbial Biochemistry, *Int. Rev. Biochem.* **21:**270–300.

Wolin, M. J., 1976, Interactions between H_2-producing and methane-producing species, in: *Microbial Formation and Utilization of Gases* (H. G. Schlegel, G. Gottschalk, and N. Pfennig, eds.), pp. 141–150, E. Goltze K. G., Gottingen.

Zaiss, U., 1981, Seasonal studies of methanogenesis and desulfurication in sediments of the River Saar, *Zentralbl. Bakteriol. Parasitenkd. Infektionskr. Hyg. Abt. 1: Orig.* C2:76–89.

Zeikus, J. G., 1981, Lignin metabolism and the carbon cycle: Polymer biosynthesis, biodegradation, and environmental recalcitrance, *Adv. Microb. Ecol.* **5:**211–243.

Zeitschel, B., 1965, Zur Sedimentation von Seston, eine produktionsbiologische Untersuchung von Sinkstoffen und Sedimentation der westlichen und mittleren Ostsee, *Kieler Meeresforsch.* **21:**55–80.

Zinder, S. H., and Brock, T. D., 1978, Methane, carbon dioxide and hydrogen sulfide production from the terminal thiol group of methionine by anaerobic lake sediments, *Appl. Environ. Microbiol.* **35:**344–352.

Zobell, C. E., 1946, *Marine Microbiology*, Chronica Botanica, Waltham, Massachusetts.

4

The Use of Microcosms for Evaluation of Interactions between Pollutants and Microorganisms

P. H. PRITCHARD AND A. W. BOURQUIN

1. Introduction

Experimental studies of the interactions within microbial communities have been a standard banner of microbial ecologists for many years. Their emphasis has been to bridge the gap between pure culture studies in the laboratory and field observations in natural ecosystems. Concern over the long-term effects of pollution on ecosystem processes has continuously challenged existing knowledge about the types, rates, and extents of these interactions, including their resistance and resilience to a large array of man-made perturbations.

The resulting interplay between research in microbial ecology and the efforts to understand the fate and effects of xenobiotics in natural systems has progressed from experimental studies with pure cultures to more complex studies using mixed cultures and communities of microorganisms (Bull, 1980; Bourquin and Pritchard, 1979). Laboratory studies with communities of microorganisms have not been particularly difficult experimentally; a large variety (but not necessarily a large number) of such studies have been generated (e.g., see Bourquin and Pritchard, 1979). This progression has been accompanied by increased concern about the "naturalness" of microbial communities studied in the laboratory. Specifically, the concern centers around the environmental

P. H. PRITCHARD and A. W. BOURQUIN • Environmental Research Laboratory, U.S. Environmental Protection Agency, Gulf Breeze, Florida 32561.

significance of information derived from mixed-culture or multispecies studies: Are the culture conditions in standard shake flasks or in completely mixed reactors containing an inoculum of some environmental sample reflective of conditions in nature? Or should greater care be exercised in developing laboratory systems that are more representative of the structural integrity of nature and that model natural interactions of microbial communities with higher trophic levels? In response, microbial ecologists, among other ecologists and toxicologists, have developed a variety of unique and practical experimental methods that address the question of how laboratory-derived data can be used in predicting potential deleterious impacts of general and specific pollution events on natural environments. An example of such a development has been the appearance of "microcosm" or "model ecosystem" studies in the areas of microbiology, ecology, and toxicology. The term *microcosm* is undoubtedly the result of a whimsical indulgence in semantics because its literal application to biology borders on absurdity. But it is a term that is used in the scientific literature with increasing frequency and has been considered a valuable descriptor of certain laboratory studies (e.g., see Giesy, 1980). Like many words of similar, nonspecific origin, however, it has a "pseudocognate" meaning; that is, each user feels that his audience shares his own intuitive definition (Salt, 1979). This situation, if not corrected by a formal definition of the word, will lead to cynicism, ambiguity, and eventually the loss of a word that could stimulate new and valuable experimental concepts.

For the purposes of this chapter, a microcosm study is defined as an attempt to bring an intact, minimally disturbed piece of an ecosystem into the laboratory for study in its natural state. It is the establishment of a physical model or simulation of *part* of the ecosystem in the laboratory within definable physical and chemical boundaries under a controlled set of experimental conditions (i.e., lighting, humidity, aeration, mixing, temperature). The design of the microcosm is a function of the question under study; the boundary conditions of a microcosm can be manipulated for the purposes of testing hypotheses and drawing inferences as to causality and mechanisms of interaction.

A microcosm is theoretically representative of the level of complexity found in the portion of the ecosystem from which it was sampled. Microcosms should therefore be self-sustaining to a large extent and should demonstrate the same homeostasis as natural ecosystems. Since this level of complexity goes beyond simple composites or mixed cultures of microorganisms (most mixed-culture studies are not necessarily microcosm studies), the principal emphasis for the microbial ecologist is on the term *simulate* or *intact,* which distinguishes a microcosm study from common mixed-culture or bacterial community studies. The attempt in a microcosm study is to avoid the arbitrary establishment of an isolated bacterial community or a mixed-culture interaction in the laboratory. Instead, an intact portion of the environment is transported to the laboratory and maintained in conditions as close to real life as possible, preserving both the physical integrity of the environmental sample and the interactions of many communities at several levels of trophic organization. Appropriate subjects for study in microcosms would be relationships between the activities of microbial communities and surfaces of animate and inanimate objects, bioturbation by invertebrate meiofauna in sediment, grazing by organisms that depend on bacteria as food sources, the exudates of plant root systems, or phytoplankton–zooplankton community dynamics. The complexity of such studies is enormous, particularly when one considers the limited experimental knowledge of the environmental factors that control the activities of isolated microbial communities. In the long term, however, a more productive approach to the study of (microbial) ecology may result (Hill and Wiegert, 1980).

Microcosms may be considered as analogues of the field; thus, studies with a microcosm are surrogates for actual field studies. As such, microcosms can be dosed with a pollutant in lieu of dosing the field, potentially circumventing the logistical, financial, and administrative problems associated with field exposures. This dosing capability is the most important attribute of and justification for research with microcosms. Microcosms should be very useful in aiding and supplementing decisions about real and potential impacts of toxic organic materials in nature.

A relatively large amount of microcosm literature (some microbiological) has recently been summarized (Witt and Gillett, 1977; Giesy, 1980; Pritchard, 1981; Levandowsky, 1977; Draggan, 1979.) A review of microcosm studies is basically a methodological consideration. Microcosms do not represent a discipline unto themselves, but instead can be applied in different ways to a wide variety of microbiological studies. Like other research methods, microcosms have certain limits and criteria for their application and for the interpretation of the results derived in their use. Much of the foundation for these applications and conceptualizations has come from microcosm studies in which bacterial processes are considered either indirectly or not at all. However, the experience gained in using the microcosm method in a variety of ecological contexts is quite relevant to microbial ecology investigations. A number of microbiological undertakings, of course, have used microcosmlike methods and concepts without formal mention of the term.

The following aspects of microcosm-related research are considered:

1. *The microcosm approach.* The specific advantages that microcosms offer over field studies are discussed.

2. *The role of microcosms in risk assessment.* Microcosms can represent an important part of the evaluation of environmental hazard by certain pollutants if used in conjunction with mixed-culture (multispecies) and pure-culture (single-species) studies.

3. *Qualitative predictors of toxic chemical fate and effects.* Microcosms can contribute to screening programs for pollutant fate and effects by providing a routine test system with ecosystem complexity.

4. *Quantitative predictors of microbial processes.* Microcosm studies help elucidate the mechanisms, kinetics, and controlling variables of microbially

mediated processes in a variety of complex ecosystem compartments, and they can help determine the environmental relevancy of data produced from simpler laboratory test methods. Specific treatments of a microcosm, judicious integration with mixed-culture studies, and interpretation of results through the use of mathematical models must be undertaken.

5. *Physical integrity of natural ecosystems*. It is argued that a proper assessment of the interactions between pollutants and microorganisms in aquatic environments must take into account the natural physical integrity of ecosystem components, e.g., sediment–water interfaces and redox gradients. A microcosm is the principal method for maintaining this integrity in the laboratory.

6. *Interaction with higher trophic levels*. It is argued that microbial responses to pollutants are tightly coupled to the interplay between microbial communities and higher organisms (e.g., grazers, phytoplankton, fungi). Microcosms provide the best laboratory means of studying these responses.

7. *Field calibration*. The successful use of a microcosm requires that it simulate a natural field situation. Various ecosystem functions and structures can be used in this calibration exercise.

8. *Precautions and potentials*. Microcosms have certain significant limitations, but their potential as a productive and informative laboratory method is considered to be excellent.

Our discussion and citations carry a certain bias in that they reflect the authors' interests in biodegradation and aquatic ecosystems (particularly estuarine).

2. History of the Microcosm Research Approach: Environmental-Risk Assessment

The need for microcosm studies originated from two important considerations: the fear of oversimplification and the problems associated with using pollutants and xenobiotics in field studies. The fear of oversimplification has been a common problem in the approach to standardized tests for assessing the potential impacts of a toxicant to an ecosystem or to a microbial community. Often, this approach has involved single species or pure cultures, and the isolation of the microorganisms from the complexities of nature has cast considerable doubt on the environmental significance of the test results. As microbiologists have recognized that the toxicity of a chemical to a pseudomonad or the degradation of a chemical by a pure culture is of little quantitative value in assessing impacts in the field, toxicologists have also recognized the need to complement single-species bioassay tests with toxicological studies on communities of organisms. The justification has been that the interactions among populations may be more sensitive to toxicants and that community studies may utlimately lead to a more quantitative estimation of ecosystem effects. The microcosm is perceived as a potential laboratory tool that could help reduce this "fear of simplicity" by permitting fate and effects studies in much more ecologically complex laboratory systems. Whether this approach will, in fact, relieve this fear depends on how microcosm results are interpreted.

Problems encountered in carrying out field studies with toxic or persistent pollutants were the second major factor behind the development of microcosms. Although scientists would like to ultimately measure the fate and effects of a pollutant by dosing a natural system (a field site), the opportunities for such experiments are extremely limited. The risk of environmental damage is a social and regulatory problem that is difficult to reconcile. In addition, the enormous problems of setting up the enclosure, mixing the chemical after dosing, and adequately sampling for information on fate and effects are well recognized (Steel and Menzel, 1978; Case, 1978).

Several attempts have been undertaken, for example, to enclose large portions of various aquatic ecosystems to study the fate and effects of such compounds as polycyclic aromatic hydrocarbons (Lee *et al.*, 1978) and heavy metals (Beers *et al.*, 1977; Davies and Gamble, 1979). These studies have used very large polyethylene bags (60,000- to 170,000-liter capacity) to trap parts of an ecosystem in the field. The MERL tanks at the University of Rhode Island (Pilson *et al.*, 1979, 1980) are a similar type of enclosure system, except the tanks are land-based, and seawater and sediment must be added to the tanks. Ecosystem parameters measured in enclosure experiments generally show good correlation with and similarity to actual field measurements. However, their cost and labor intensity restrict their application to a variety of pollutants. It is also not clear how much of the information gained from enclosure experiments might have been obtained in smaller and more convenient systems in the laboratory.

If a section of the environment could be brought into the laboratory (as a microcosm) and shown to behave as in the field, then it would offer a more practical system to dose with a toxicant. It would provide a number of significant advantages over a field study. First, the recognition of a particular toxic effect or fate phenomenon could be greatly enhanced, since natural variations in functional and structural parameters caused by seasonal and climatic events could be minimized. This assumes that a microcosm operated under a standard set of laboratory conditions will still serve as a reasonable analogue or surrogate for the field. It also assumes that results obtained from the microcosm could be extrapolated to the field over large spans of space and time. Various studies indicate that these assumptions are correct (Harte *et al.*, 1980; Hill and Wiegert, 1980; Schindler *et al.*, 1980). Second, the small size of the microcosm permits replication, simplified dosing mechanics, control over imports and exports from the system, and adequate mixing, all factors that are difficult to control in the field. The microcosm gives the researcher the opportunity to vary environmental conditions to simulate a large array of possible perturbations to the system. These options are not available in field studies. If the proper concepts and logic can be developed to extrapolate microcosm results to the field, then the methodology has great promise for understanding the interactions of natural communities, both at the microbial level and at higher trophic levels.

3. Role of Microcosms in Environmental-Risk Assessment

Information from laboratory tests can be used to evaluate the behavior of a chemical in the environment in a *qualitative* or *screening* type assessment. Toxicity testing, for example, involves a determination of the concentration at which each test chemical will elicit a specific effect—such as mortality, decreased growth rate, or decreased metabolic activity—using the same highly standardized bioassay test. Biodegradability testing involves the measurement of the extent to or rate at which a chemical is degraded, using standardized measures of biological oxygen demand, mineralization, or loss of parent compound, and a standard inoculum such as sewage sludge. In both cases, risk assessment is based on a relative comparison of test results in which the test information from one chemical is ranked within a data base for a large array of previously tested chemicals by a standardized method. The test methods do not necessarily produce environmentally realistic data, but rather "worst-case" situations. The emphasis is on consistency of results and methods of standardization. Decisions on which chemicals are considered a potential hazard to the environment are based on scientific prudence, common sense, and consensus judgments.

The second use of laboratory tests in evaluating the behavior of a chemical in the environment is a *quantitative* assessment, or a determination of waste assimilative capacity [the potential of any environment to accept pollution without adverse effects (Cairns, 1981)]. An assessment of this type, of course, requires accurate measures of the rates and dynamics of both fate and effects processes, including their interactions, and a determination of how toxic chemicals will quantitatively alter these processes. It requires answers to questions such as: (1) How much of a reduction in microbial population density, activity, or diversity can be tolerated before permanent damage (based on some carefully qualified time scales) occurs? (2) How much of a response is required before ecosystem balance is upset? (3) How much of a difference in biodegradation rates of a chemical is allowed before the chemical is considered persistent in one environmental area and not in another?

Microcosms can be potentially used in both screening programs and waste assimilative capacity determinations. Figure 1 is a diagrammatic representation of how information from laboratory fate and effects tests (including microcosms) is currently applied to relative (screening) assessments and to waste

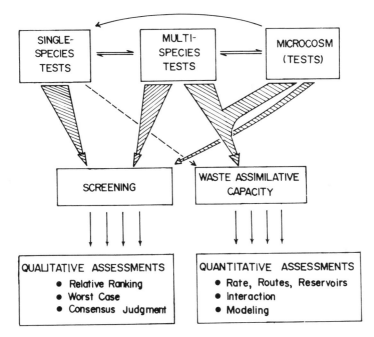

Figure 1. Schematic diagram of possible information flow from laboratory test systems to risk assessments. Differences in tests are explained in the text. Hatched arrows represent current and projected flow of information.

assimilative capacity determinations. Fate and effects testing is divided into three levels of organization: (1) single-species tests, which use pure cultures of bacteria and other organisms; (2) multispecies or mixed-culture tests, which use inocula from the field in an experimentally convenient manner; and (3) microcosm tests, which use intact field samples from the environment.

Mixed-culture tests ignore any controlling factors that might result from (1) taking intact field samples and maintaining them in the laboratory under natural field conditions, or (2) isolating a bacterial community from interactions with other communities (plant, animal, and bacterial) or trophic structures. These tests generally have a minimal number of physical components. The emphasis in complexity is confined to mixtures of species or isolated communities and does not usually include relationships with other components in the ecosystem.

Microcosm tests, on the other hand, are basically more rigorous mixed-culture tests, involving a number of physical components and involving studies of environmental samples in the laboratory, where the natural physical integrity of the field sample is maximized and isolation from other ecosystem components is minimized.

3.1. Single-Species Tests

Single-species tests primarily support relative assessments or screening programs. This is particularly true of the single-species bioassays used in many areas of toxicology. Little of this information can be applied to quantitative assessments. In many cases, however, the response of a single species to a particular chemical may provide invaluable mechanistic information that can be applied to the interpretation of a more complex experimental system. Pure cultures of bacteria can, of course, be used to examine in detail the toxicity of certain chemicals, for example, the effects of chlordane on *Bacillus, Streptococcus,* and *Mycobacterium* (Widdus *et al.,* 1971). They can be used to examine the mechanisms by which bacteria degrade specific chemicals, for example, mechanisms of cometabolism and biodegradation of DDT (Pfaender and Alexander, 1972). Although quantitative information on fate and effects processes (such as rates of biodegradation or extents of metabolic inhibition) can be obtained from single-species tests, it has been pointed out (Bull, 1980; Tempest, 1970; Levin, 1982) that the measurements cannot easily be applied to natural field situations because the test organisms are extensively laboratory-acclimated and experimental conditions provide for optimized growth and survival, a condition unlikely to be found in the field. This is why, in many respects, "worst-case" interpretations have been invoked.

3.2. Multispecies Tests: Qualitative Assessments

Multispecies tests, that is, the use of mixed cultures or communities of microorganisms for a testing protocol, have been used extensively to assess the potential fate and effects of xenobiotics. These tests support both qualitative (screening) and quantitative assessments (waste assimilative capacity) (Fig. 1). Screening biodegradation tests using microbial communities, for example, commonly follow the disappearance or metabolism of a chemical in samples of water, sediment, soil, or sewage taken directly from the field and incubated in various types of laboratory containers (Howard *et al.,* 1978; Gerike and Fischer, 1979). Insight into modes of biodegradation in actual environmental media that could not have been gained from pure-culture studies is often obtained from multispecies tests. Knowledge of the ability of sediment-associated bacteria to degrade chemicals faster than water-associated bacteria, for example, has come from such studies (Lee and Ryan, 1979; Juengst and Alexander, 1975; Simsiman and Chesters, 1976; Graetz *et al.,* 1970).

Similar enhancement of knowledge (relative to pure-culture studies) about the toxic effects of chemicals on microorganisms and the biogeochemical processes they mediate has also been obtained in mixed-culture studies (Colwell, 1978). Effects of toxic chemicals on heterotrophic activity in aquatic systems (for example, Sayler *et al.,* 1979; Orndorff and Colwell, 1980; Goulder *et al.,* 1980; Albright and Wilson, 1974), on nitrogen cycling in estuaries (R. D.

Jones and Hood, 1980), and on a variety of microbial activities in soil (Tu, 1980; Atlas *et al.*, 1978; J. R. Anderson, 1978) are representative of the increased amount of environmentally relevant information available from multispecies tests as opposed to single-species tests.

The information from a mixed-culture test can be evaluated on the same relative basis as for pure-culture tests. In this case, the environmental risk of a chemical is determined by relative comparisons to chemicals that have a field-documented history of persistence or toxicity in nature. A compound as persistent as Kepone, for example, would present considerable environmental risk. However, many chemicals fall into the "gray areas" of degradability and toxicity and thus are much more difficult to evaluate in terms of potential environmental contamination by making relative comparisons of data.

3.3. Multispecies Tests: Quantitative Assessments

Mixed-culture-test information can also be used for quantitative assessment if: (1) it involves the determination of rates, routes, and extents; and (2) it can be related to or integrated with some larger picture of the whole (the ecosystem). The necessity for rates and rate constants results from obligatory use of ecosystem mathematical models for predictions of waste assimilative capacity (Cairns, 1981; Lassiter *et al.*, 1978; Hill, 1979). Integration into the whole is necessary because mixed-culture tests, by the very nature of their pragmatic design and operation, isolate microbial communities from interactions with other communities, with other trophic levels, and with the physical structure of whole ecosystems. Consequently, although rate information can be derived easily from mixed-culture tests, it is basically meaningless by itself until its relationship to the whole or until its coupling to ecosystem function is established and interpreted. It is the author's opinion that this relationship to the whole is the key to extrapolating laboratory data to the field and the foundation for using microcosms in fate and effect research. As Heath (1979) has stated, a(n) (eco)system is an array of components interacting cooperatively to form a complex unit through which energy, matter, and information may flow. The behavior of the individual components of a system is constrained and coordinated through the network design of the system. Since some of these properties of the system arise not from the components themselves but from the specific set of interactions within the system, it is not possible to characterize a system only from a knowledge of the component parts. Further, only by coupling a holistic investigation of the intact unit to reductionist studies (mixed-culture studies) can we hope to understand the natural functions of ecosystems (Heath, 1979). There are many aspects of microbial ecology in which the knowledge and experimentation with a state-of-the-whole is crucial for understanding not only the mechanisms of microbial interactions but also the rates at which they occur.

A microcosm is an excellent method for studying the state-of-the-whole in the laboratory. Figure 1 emphasizes that in order to apply information from multispecies tests to waste assimilative capacity determinations (that is, to extrapolate laboratory data to the field), studies on the fate and effects of chemicals in microcosms must also be considered. Thus, the role of a microcosm is verification; it basically tells a researcher how to apply information from mixed-culture studies to complex real-world situations. It is assumed that the microcosm, as a surrogate for the field, is being used because exposure of the field to a chemical pollutant is unfeasible; a field study would be the ultimate verification if it were practical to implement.

3.4. Application of Mathematical Models

Predictions about the waste assimilative capacity of an ecosystem require, in many cases, the use of mathematical models. Models are the only means of evaluating the behavior (fate or effect) of a pollutant when several processes and their controlling variables operate simultaneously and perhaps interact to varying degrees. For example, a good quantitative assessment of the fate of a chemical should collaterally analyze information on the kinetics of each mitigating process (hydrolysis, biodegradation, sorption), on the inputs (time variable, point and nonpoint) of the chemical into the system, on the physical movement and compartmentalization of the chemical (sorption, sediment transport, mixing, diffusion), and on specific environment (pH, salinity, dissolved oxygen) and meteorological (temperature, runoff) variables associated with a particular ecosystem. The most efficient mechanism for such an analysis is a mathematical description of the total effect that this information will have on the concentration of a chemical through some finite interval of space and time. Such a description usually takes the form of a large set of coupled nonlinear differential equations that can be generally solved by numerical integration.

This exercise is complicated and expensive. In reaching a compromise between resolution and economy of resources, modelers frequently divide the exercise into two activities: the development of *process* or *analytical models* and the development of *system* or *ecological simulation models*. The former is always a subset of the latter. The system model concentrates on environmental dynamics or on those forcing functions over which the ecosystem has little control, i.e., runoff from the watershed, insolation, hydrodynamics, sediment scour, and deposition. These are generally environmentally derived parameters that often use the distributional history of a conservative organic chemical (such as Kepone or PCBs) as a calibration source. Each function contributes to the output of the total system by interacting in specific ways with a series of statevariables or components described for that ecosystem. The simulation modeling approach, therefore, is basically a tool for, first, systematizing knowledge about the complex networks involved in the fate and effects of a pollutant in an ecosystem and, second, for logically and reproducibly deriving implications about potential environmental hazard or assimilative capacity (Baughman and Burns, 1980). At each stage in the model development, approximations infer some assumption about the real world. Recognition and understanding of the assumptions are the key to evaluating the results of the model.

The assumptions permit the construction of kinetic expressions for the basic processes within ecosystem components that describe, quantitatively, the dynamics, controlling variables, and interplay with various other state variables. For the fate of a chemical, these expressions describe biodegradation, hydrolysis, photolysis, and so forth; for microbial toxicity studies, these expressions would describe population growth, nutrient cycling, mechanisms of biological deficit compensation, and so forth. The development of these kinetic expressions is referred to as process or analytical modeling (Kremer, 1979; Baughman and Burns, 1980). These models couple laboratory-derived process information with the field-derived parameters of the systems simulation model. If information about these processes is insufficient to develop the appropriate kinetic expressions, an experimental program must be implemented to provide that information. It is the opinion of the authors that the success of such a program lies in the quantitative use of microcosms. Microcosms provide a means of testing and verifying the theoretical constructs and assumptions used in the development of the kinetic expressions under conditions similar to those in the field (Fig. 2). Further, any *modifications* in the assumptions and the theory that are conceived as potential improvements in our predictive capabilities about the fate and effects of pollutants can be tested in a similar manner

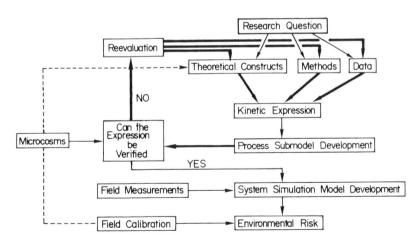

Figure 2. Flow diagram depicting the use of microcosms as verification tools. The basic iteration involves verifying the conceptual and experimental basis of the kinetic expressions using complex natural situations such as microcosms. Once the kinetic expressions are verified, they can be used in ecosystem simulation models.

(Fig. 2). As a consequence, the modeler has sufficient justification to couple the kinetic expressions of the process model to the system simulation model. It should be pointed out, however, that the basic information for the kinetic expressions (i.e., rate constants) will come from standard laboratory tests and not from the microcosm (Fig. 2). The iterative-type process diagram in Fig. 2 is, in the authors' opinion, the most logical and rational way to extrapolate laboratory data to the field.

4. Typical Microcosm Design Features

The definition of and approach to using microcosms having been presented, it is appropriate, at this point, to physically describe a "typical" microcosm for those unfamiliar with the array of design features commonly employed. Microcosm design is, of course, closely related to the experimental question being considered, the environmental site being modeled, and the investigator's preference.

4.1. Size

Microcosm vessels generally vary in volume from several hundred milliliters to several hundred liters. For aquatic studies, Pritchard et al. (1979) used a 3.5-cm-diameter glass tube containing 200 ml water, in contrast to Perez et al. (1977) and Harte et al. (1980), who used glass tanks containing 150 and 200 liters, respectively. Quite commonly, 20-gallon aquaria have been used with considerable success (Giddings and Eddleman, 1978; Sikka and Rice, 1973). Long fiberglass troughs, approximately 30 cm square and 6 to 12 m long (S. R. Hansen and Garton, 1982; McIntire, 1965), have been used to model streams. Terrestrial systems also span a large size range from 40-cm^3 glass or plastic columns (Draggan, 1977; Jackson et al., 1977) to large glass boxes with volumes in excess of 800 cm^3 (Nash and Beall, 1977; Gillett and Gile, 1976).

The shape and size of the vessel are dictated by the level of trophic structure to be studied (incorporating plants, for example, will necessitate larger containers), scaling factors (modeling natural sediment surface area/water volume ratio), and sampling intensity, particularly when sediment or soil is to be removed from the microcosm for analysis. In some studies not directly carried out for microbiological purposes, consistent responses of certain measured parameters could be obtained only with larger microcosms (100–200 liters) (Harte et al., 1980; Giddings and Eddleman, 1977). Microbially mediated processes, for the most part, can be readily studied in smaller microcosms.

4.2. Inoculation

The most important design feature of a microcosm is the inoculum. Ideally, an intact piece of the environment, as suggested earlier, should be placed in the microcosm vessel as undisturbed as possible. A system concerned with microbial activities only in the water column, where interactions with sediment are considered insignificant, probably requires no special care in the transfer of the inoculum, except perhaps to minimize the inevitable bottle effects. It is questionable whether microcosm studies should even be considered in water-column studies. Inocula from sediment and soil are probably best obtained as intact cores. Numerous microbiological studies with intact cores have produced evidence that these cores adequately represent natural conditions in the field (Graetz et al., 1973; Pamatmat, 1971; Cappenberg, 1974; J. G. Jones, 1979; Craib, 1965; Van Voris et al., 1980; Twinch and Breen, 1981; Henriksen et al., 1981). The corer itself, depending on the application, can be converted into the microcosm (Pritchard et al., 1979; Van Voris et al., 1980), or the core can be extruded intact into a separate microcosm container. Relatively large cores that can encompass rooted plants and a variety of invertebrate animals are also possible.

Where coring is not possible, the container can be filled with natural sediment or soil and returned to the field for acclimation and the recovery of its original undisturbed state. Redox gradients, which are easily disturbed by sampling and may be critical to certain microbial processes, can be reestablished in this manner (Flint et al., 1978). The container can then be removed from the field and transported to the laboratory.

A final alternative is to collect environmental material without concern for natural physical integrity and then allow the material in the microcosm to reorganize itself into a miniature, albeit unique, ecosystem. It is assumed that most of the processes and interactions typically seen in the field will reestablish themselves in the microcosm if given enough incubation time and if the proper inputs are allowed (Lassiter, 1979). Several stream microcosms have been established in this manner by colonizing clean rocks and sand with natural stream water delivered continuously over long periods (McIntire, 1965). The adaptability and functional redundancy in microbial communities make this approach feasible. Lassiter (1979) suggested that an understanding of the interworkings of these microcosms will aid our understanding of natural processes and their interactions, since complexities are comparable. Regardless, the inoculation process is time-consuming, and the criteria necessary to establish the maturity or equilibrium of these ecosystems are not clearly specified. Coring methods appear to provide adequate simulations of the field with a fraction of the incubation time and therefore would tend to supplant this latter inoculation method.

4.3. Mixing

Microcosm studies used to demonstrate the response of bacterial populations to pollutants generally require some type of dispersive action to assure uniform distribution of the pollutant. Aquatic microcosms are generally operated with a completely mixed water column to simplify sampling and interpretation of results. Such a mixed state is mechanically quite difficult to obtain, particularly when light resuspendable sediment is present. When replicate microcosms are required, mixing must be equal in all systems. The biodegradation of certain chemicals is, in some cases, closely linked to the presence of sediment (Pritchard et al., 1979; Graetz et al., 1970; Lee and Ryan, 1979); consequently, the degree of mixing will control the degree to which the chemical is physically exposed to the sediments and their associated microorganisms, and the extent of biodegradation.

Mixing is generally accomplished by rotating paddles, propellers, or specially shaped rods. A small motor drive attached to each microcosm is most efficient. Geared, preset-speed motors are preferred to variable-speed motors since, with the latter, constant low speeds are often difficult to maintain. Moderate bubbling with air can provide mixing in small microcosms (Pritchard et al., 1979). Mixing by recirculation through magnetically driven or peristaltic pumps is also very efficient but adds considerable surface area, which may lead to additional wall growth and possible adsorption of the pollutant. Application of a pollutant to terrestrial microcosms is generally accomplished through simulated rain events (Van Voris et al., 1980; Gillett and Gile, 1976; Nash and Beall, 1977). Reasonably uniform distributions have been obtained.

4.4. Other Accessories and Operations

Again, depending on the application, various other microcosm design features must be considered. Temperature is usually controlled by water baths or constant-temperature rooms. Lighting is often provided by a bank of 40 W cool-white fluorescent lights, which do not present heat problems and have spectral and intensity qualities that are adequate to support plankton and plant growth. Various types of monitoring equipment are often incorporated in the microcosms and most commonly provide continuous measurements of dissolved oxygen, pH, carbon dioxide, and temperature.

The duration of a microcosm experiment can vary from hours to months, depending on the particular application. Often, pretest incubations are carried out to establish base-line information and sustainability in the laboratory. Pollutants can be added to a microcosm in single doses or by continuous dosing. In the latter case, the microcosm can be considered a complex extension of continuous-culture systems. Various types of inputs are often required for long-term microcosm experiments; aquatic microcosms frequently receive untreated water as a continual renewal of nutrients and microorganisms.

5. Qualitative Approach for Microcosms

In much the same way as a mixed-culture or multispecies test can be dosed with a chemical and the results evaluated on a relative basis, a microcosm, with its additional complexity of maintaining natural physical integrity and higher trophic level interactions in the environmental sample, can be dosed with a variety of chemicals and the results evaluated in a qualitative or relative manner. The actual relationship of the microcosm results to the field is not necessarily the principal point of concern. Instead, reproducibility of both the complexity of the test and the relative responses of the test to a pollutant are the key elements.

5.1. Ecosystem-Level Responses

A major assumption in the development of screening microcosms is that through the use of these more complex test systems, new or subtle responses of microbial communities, or of ecosystems in general, to a pollutant will be observable. If this is true, then it becomes a question of which parameters or behaviors can be measured. Ideally, measurements of general ecosystem behavior, i.e., the summed results of many processes operating simultaneously, should be considered. For example, ecosystems can be thought of as microbially based biogeochemical units that have distinct measurable properties (Schindler et al., 1980). These properties may include community respiration, nutrient cycling, oxidation–reduction gradients, organic matter transformation, and primary and secondary production. All these properties are "ecosystem-level" behaviors that reflect the integration of many internal processes and components and their interactions. Ecosystem-level behaviors presumably cannot be inferred or predicted from measurements of isolated components (single species or mixed cultures) in the laboratory (Weiss, 1971). Many of these behaviors are known to reflect a significant degree of homeostasis in an ecosystem, the essence of which is controlled through the tight cycling of certain essential elements by microorganisms and their interactions with higher trophic levels (Fry, 1982). Ecosystem homeostasis is responsible for the maintenance of standing biomass and for the persistence of complex natural systems. It is the interference of a pollutant with the biogeochemical activities of microbial communities that can cause potentially drastic environmental consequences.

Microcosms, as analogues of the field, should also be biogeochemical units showing the same ecosystem-level behavior and homeostasis. A screening program can therefore be developed in which toxic chemicals are administered to a particular microcosm system and the response of the ecosystem-level behavior (fate and effects) is measured and compared (to establish risk) on a relative basis, much like the assessment procedures used with bioassay and biodegradability information. Although this approach is still somewhat speculative, some success has been reported (Harris, 1980).

5.2. Screening Microcosms

The original attempt at screening chemicals for biodegradability and bioaccumulation in complex laboratory systems was pioneered by Metcalf and colleagues (Metcalf *et al.,* 1971; L. K. Cole *et al.,* 1976). The systems they used were not microcosms by our definition, but synthetic communities established by combining specific laboratory-reared organisms (alga, *Oedogonium;* mosquito larva, *Culex;* water flea, *Daphnia;* snail, *Physa;* fish, *Gambusia*) in an aquarium containing a terrestrial section (sand planted with *Sorghum*) and an aquatic section (reference pond water). The radiolabeled test chemical was applied to the plants (to mimic field application), and the distribution of the parent compound and its degradation products through the system was determined following the consumption and defecation of the plant material by a saltmarsh caterpillar. By comparing the relative distribution of the chemicals, these workers formulated a "biodegradability index" (the ratio of solvent-unextractable material or polar products found in water or animal tissue to the amount of solvent-extractable material) (Metcalf *et al.,* 1973). Although the application of the data generated from these screening systems (which include an impressive number of chemicals) to regulatory decision-making has been questioned (Pritchard, 1981; Branson, 1978; Witt and Gillett, 1977; Witherspoon *et al.,* 1976), the Metcalf effort is significant as the first attempt to apply "ecosystem complexity" to screening programs. It also stimulated similar experimentation with somewhat more complex synthetic community systems and with emphasis on different types of model communities, namely, aquatic (Isensee *et al.,* 1973) and terrestrial (Gillett and Gile, 1976; Nash *et al.,* 1977; Lichtenstein *et al.,* 1978). The synthetic community approach, however, does not consider system-level properties because community interactions are, by the design of the tests, kept at a minimum.

Many of the original microcosm studies of Beyers (1964), Abbott (1966), Cooke (1967), and McIntire (1965) showed that complex natural communities could be maintained in the laboratory over long periods of time and that these systems, as integrated units, had measurable system-level properties (mainly functional parameters) and homeostatic responses that were not only reproducible but also to some extent mimicked similar response measurements in the field (Cooke, 1967; Schindler *et al.,* 1980; Odum, 1969).

Relatively little further development of these studies for screening purposes has resulted, particularly as applied to microbial communities. Hammonds (1981) and Harris (1980) have reviewed laboratory studies that potentially could be used for screening microcosms by the U.S. Environmental Protection Agency. Two illustrative examples are in order.

Giddings and Eddleman (1978) (also see Harris, 1980) have shown that small ponds can be modeled in microcosms for potential screening purposes. This was accomplished in laboratory aquaria by establishing an *Elodea* plant community in sediment samples taken from the ponds. Populations of microorganisms, diatoms, unicellular algae, protozoa, benthic invertebrates, some insect larvae, and the snail *Physa* developed in the microcosm into a balanced community within about 6–10 weeks. Thus, the microcosm criteria of modeling physical integrity and maintaining couplings with higher trophic levels were considered. Community metabolism [ratio of primary production (P) to respiration (R)] and nutrient cycling were used as the principal ecosystem-level measurements. Distinct similarities in P/R ratios, concentrations of dissolved organic carbon, phosphate, nitrate, and ammonium concentrations, and plant and animal community structure between pond and microcosm were observed for an incubation period of 170 days. Coefficients of variation among replicate microcosms were generally below 20%, which is considered acceptable for a screening system (Harris, 1980). This microcosm system can be dosed with a variety of toxic chemicals and the fate (biodegradation and distribution) and effects (on nutrient cycling or organic matter turnover) of the chemicals observed on a relative basis. As Giddings stated (see Hammonds, 1981), the rankings, not the observed effects, constitute the output of this experimental design. Microcosms subjected to stress with arsenic (Giddings and Eddleman, 1978; Harris, 1980) showed significant perturbation in P/R ratios and pH, and drastic increases in inorganic nutrients and total organic carbon (Fig. 3). The fate of anthracene in these microcosms has also been determined (Giddings *et al.*, 1979). These fate and effects studies formed the basis for additional screening of other xenobiotics. The studies are an excellent example of the screening microcosm approach that can be used to study responses of microbial communities to toxicants.

A soil-core microcosm has also been developed for screening chemicals in terrestrial systems (O'Neill *et al.*, 1977; Jackson, *et al.*, 1977, 1979; Draggan, 1977). Again, microbial geochemical processes form the backbone of their toxic-effect responses. O'Neill *et al.* (1977) proposed that the rate at which nutrients are leached from soil could be used as a monitoring point to detect pollutant-induced changes in ecosystem behavior because nutrient flux in terrestrial ecosystems results from the "functional synchrony of autotrophs, heterotrophs, and soil organic matter." Similarly, Van Voris *et al.* (1980) suggested that CO_2 efflux from the respiration of soil microorganisms could also be used as an ecosystem-level measure of pollutant stress in soil. Support for

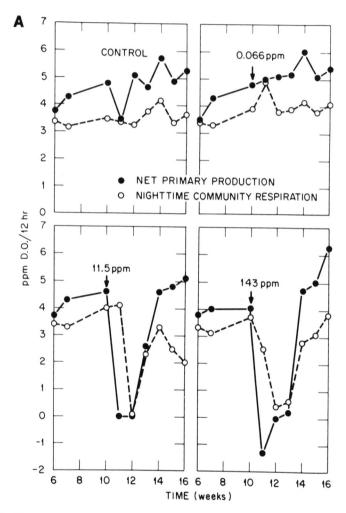

Figure 3. Net primary production and nighttime community respiration (A) and production/ respiration ratios (B) in microcosms at three arsenic levels. (O) Nighttime community respiration; (●) net production. Each point represents the mean of three microcosms. Arsenic was added, in the concentrations indicated, immediately following *P* and *R* measurement. After Giddings and Eddleman (1978).

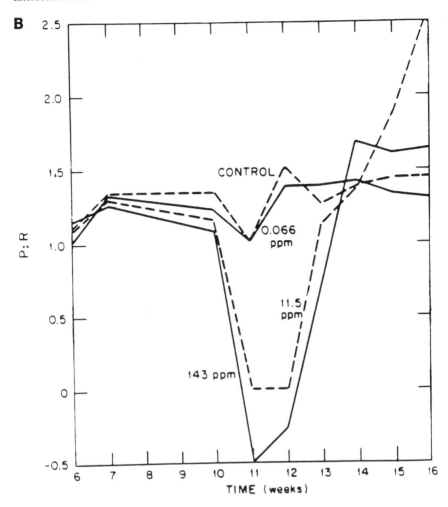

Figure 3. *(continued)*

these suggestions has come from the development and testing of microcosms in which intact soil cores (containing autotrophs) are brought into the laboratory and maintained in plastic columns designed for the continuous monitoring of CO_2 and for the continuous collection of leachable nutrients resulting from simulated rain events. A diagram of the microcosm system is shown in Fig. 4. Arsenic and smelter wastes were found to disrupt nutrient cycling and the concentration of Ca^{2+}, NO_3^-, NH_4^+, and PO_4^{3-} leached from the microcosm. These changes were a more sensitive measure of effects than those in common pop-

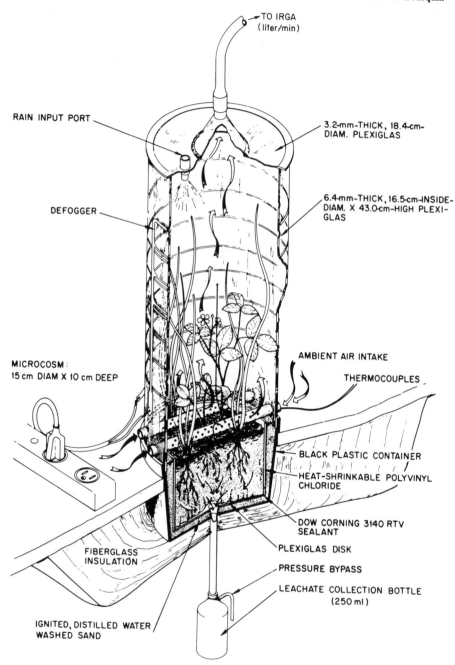

Figure 4. Cross section of encased and cuvette-covered soil-core microcosm showing the two ecosystem-level monitoring points of CO_2 and nutrient export. (IRGA) Infrared gas analysis. After Van Voris et al. (1980).

15-cm diam EXCISED MICROCOSMS

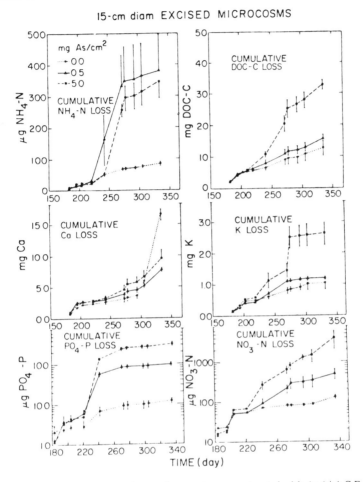

Figure 5. Cumulative nutrient loss from soil core microcosms treated with As (± 1 S.E., $N = 3$). After Jackson *et al.* (1979).

ulation parameters, such as ATP concentrations, bacterial density, and plant growth (O'Neill *et al.*, 1977). An example of this effect is shown in Fig. 5.

The characteristics of these microcosms agree well with the tenets suggested for microcosms; in fact, intact soil cores gave a more sensitive toxic response than homogenized soil (Draggan, 1977; Jackson *et al.*, 1979). Toxicant distribution and extractable nutrient concentrations during treatment of the soil cores were comparable to similar results from dosed field plots (Jackson *et al.*, 1979). The functional complexity and ecosystem stability of these soil systems have also been demonstrated at the ecosystem level of organization (Van Voris *et al.*, 1980).

Again, it is important to reemphasize that in the use of these soil-core microcosms to screen chemicals for toxic effects, little, if any, information is provided on the mechanisms and specific microbial processes involved in the leaching and respiration phenomena. The concern instead is to rank chemicals for their effects (and fate) in a simple, expedient, and environmentally relevant manner and then draw an arbitrary threshold that will separate the potentially harmful chemicals from those that are harmless.

5.3. Criteria for the Development of Screening Microcosms

The decision regarding when to use a screening microcosm as opposed to mixed cultures studied in a shake flask is often based on nebulous criteria. As a result, microcosms, along with their additional expense and manpower requirements, are often used for screening purposes when in fact there is little justification for their use. The three factors discussed below should therefore be considered in the development and use of screening microcosms.

5.3.1. Selection of Most Simple Test

Care should be taken to assure that the conclusion or risk assessments obtained from a screening microcosm cannot be obtained in some simpler way. The mere effort of generating information from the microcosm often overshadows careful evaluation of the results and promotes insensitivity to the question of obtaining the same information from a simpler system. For example, in studies designed to determine effects on the microbiota in a microcosm, a toxic response can frequently be attributed to a specific group of microorganisms. By development of a bioassay with this group of microorganisms, the same protection of the ecosystem (worst-case analysis) can be realized through a simpler test.

In a similar way, ranking the biodegradability of chemicals with a biological oxygen demand test is simpler than using a microcosm, particularly if the dependency of persistence on site-specific considerations (which a microcosm might otherwise be designed to handle) is of no interest. Studies in which the fate of a pollutant in a microcosm has been reported are increasing in number, yet it is not clear how the information will be useful in an environmental risk assessment relative to other existing information. For example, Shaw and Hopke (1975) found that the herbicide diquat accumulated in the sediments of a microcosm. Virtanen et al. (1982) reported similar results for DDT. This information, in both cases, could have been easily obtained by using a shake flash and a sediment–water slurry. The extent of sorption to sediments in the microcosms has no bearing on sorption in the field, since structural and functional relationships between the microcosm and the field are unknown. The dynamics of diquat and DDT in the microcosms cannot be gauged relative to other chemicals because the fate of additional chemicals has not been studied in microcosms. Finally, the measurement of a pollutant's fate in a complex laboratory system is basically valueless unless it can be shown that the results are significantly different from those generated in simpler tests. In other cases, the abiotic degradation of carbaryl (Liu *et al.*, 1981), the effect of oxygenation conditions on the degradation of pentachlorophenol (Boyle *et al.*, 1980; Liu *et al.*, 1981), and the effect of input conditions on the degradation of trifluralin (Yockim *et al.*, 1980) are results from complex microcosm experiments that might have been easily obtained from simpler mixed-culture flask studies.

To be useful, the screening microcosm must provide insight into phenomena or events that *cannot* be readily examined or demonstrated in other previously established mixed-culture tests. The soil-core microcosm developed by Van Voris *et al.* (1980) and Jackson *et al.* (1977) represents a good example of a unique screening microcosm. No other existing laboratory test provides information on the sensitivity of microbial geochemical processes to toxicants under the complex situations found in natural terrestrial environment. At some future time, however, the sensitive factor or factors that control the nutrient cycling and leaching process may be identified, thus providing the basis for a new bioassay that could then supplant the microcosm approach.

5.3.2. Uniqueness of Results

Part of the concept of water quality criteria and its use of screening data is that maximum acceptable toxicant concentration values are based on the assumption that if the most sensitive species are protected, all other species will also be protected. If the results from a screening microcosm are to be unique relative to existing simpler tests, and if they are to be evaluated in the same qualitative manner as single-species or pure-culture information, then they must indicate a greater sensitivity to a toxicant (effect at a lower concentration of test chemical) or greater resolution of fate processes than results from other well-established tests. As a hypothetical example, it would be of no consequence that 100 μg/liter of a toxicant reduces the metabolism of a microbial community on the surface of aquatic macrophytes (which was previously established as a critical ecosystem function) if free-swimming aquatic bacteria are killed at 1.0 μg/liter. Development of water quality standards to protect the most sensitive microbially mediated processes would thus use a maximum acceptable toxicant concentration of something below 1.0 μg/liter. Screening for new effects responses should therefore produce tests that show greater sensitivity to a toxicant than other established screening tools.

5.3.3. Reproducibility and Repeatability

The measurement of microbial community responses in a screening microcosm must possess a degree of reproducibility and repeatability that will allow toxic effects of a chemical to be distinguished from background noise or allow the fate of a chemical to be consistently ranked among a group of previously tested chemicals. The measurement of these responses must show good replication among microcosms within a given experiment and good reproducibility of microcosm results from experiment to experiment. This problem has been specifically addressed in several microcosm studies (Harris, 1980; Harte *et al.,* 1980; Stay, 1980; Brockway *et al.,* 1979; Isensee, 1976). In general, replicate microcosms have coefficients of variation from 10 to 30%, although, depending on which parameters are measured, variations can be substantially higher. Functional behaviors of microcosms, such as respiration, production, nutrient cycling, and litter decomposition, typically show lower variations among and within microcosm experiments than structural properties, such as population composition, species diversity, and community biomass (Harris, 1980). Functional parameters also seem to diverge less with long-term incubation (>30 days) of microcosms. Good reproducibility in the distribution of arsenic in water, sediments, and plants (average coefficient of variation was 26%, range was 4–45%) has been demonstrated in a pond microcosm system (Giddings and Eddleman, 1977). Nutrient cycling in the same microcosm system showed, in most cases, average coefficients of variation among 3, 6, and 12 replicates of less than 20% (Harris, 1980). Concentrations of NO_3^-, NH_4^+, PO_4^{3-}, total P, and total dissolved organic carbon (DOC) in the water columns of 12 microcosm replicates are shown in Fig. 6. Brockway *et al.* (1979) also had similar success in measuring DOC in pond microcosms. Other variables such as pH, dissolved oxygen, and conductivity have also shown low coefficients of variation (Harris, 1980; Brockway *et al.,* 1979). Many of these functional properties are microbially based and system-level in nature. They could therefore serve as the basis for toxicity testing in a screening microcosm if the effect is greater than the experimental coefficients of variation. Factors such as increased microcosm size (70 to 700 liters) and increased pretest laboratory acclimation (0 to 30 days) (Giddings and Eddleman, 1977; Harris, 1980) also decreased coefficients of variation. Harte *et al.* (1980), using microcosms that model the pelagic components of lakes, have shown that the decomposition response of microbial communities, as measured by the production of ammonia, to increasing concentrations of added detritus (up to 300% increase over ambient concentrations) was linear and essentially of the same magnitude in each of three replicate microcosms.

Replication and reproducibility of screening microcosms are therefore feasible in relation to responses of microbial communities. However, great care should be taken to examine variability, since microcosm design and operational features of the microcosm can greatly affect the results. It is important to further point out that once a particular biological response in a microcosm is established for screening purposes, the same system and its protocols must be utilized for each new chemical tested. Otherwise, the separation of a response to a particular chemical from background noise may be greatly confounded.

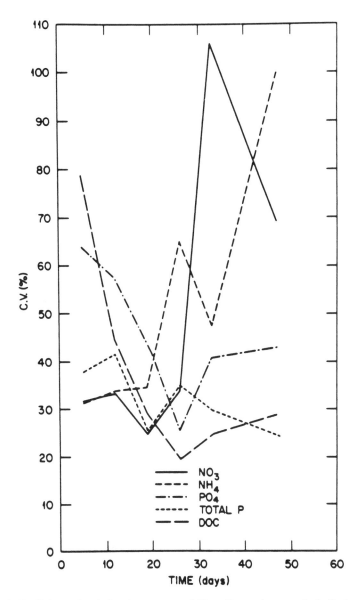

Figure 6. Coefficients of variation (percentages of 12 replicate microcosms) of nitrate, ammonium, phosphate, total phosphorus, and dissolved organic carbon (DOC) concentration in 12 replicate 70-liter pond microcosms. Microcosms contained pond water, sediment, and a macrophyte community. After Harris (1980).

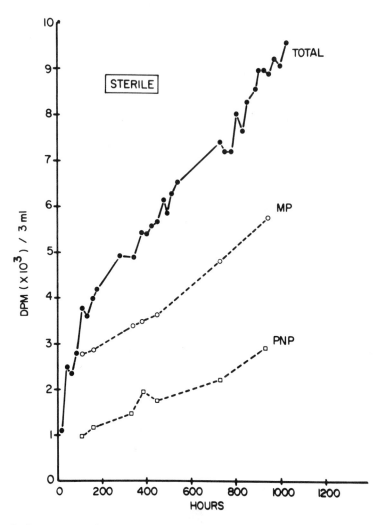

Figure 7. Concentration of total radioactivity (TOTAL), methyl parathion (MP), *p*-nitrophenol (PNP), and unextractable radioactivity (POLAR) in water overlying sediment in sterile and nonsterile microcosm systems. Unextractable radioactivity refers to radioactive products remaining in a water sample after two extractions with methylene chloride under acidic conditions. The flow rate of seawater into the microcosms was 0.35 liter/day, giving a retention time of 48 hr. The inflow concentration of radioactivity was approximately 12,000 cpm/3 ml or 20 μg/liter methyl parathion. Sterile microcosms were initially sterilized by adding formalin to a concentration of 2% (vol./vol.). The presence of 0.1% formalin in the inflowing toxicant solution (sterile reservoir) was sufficient to maintain sterility in the reaction vessel even with continuous addition of unsterilized seawater. After Pritchard *et al.* (1983b).

6. Quantitative Approach to Using Microcosms for Environmental-Fate Analysis

The role of microcosms in quantitative environmental-risk analysis was discussed in Section 3.3. Studies on the fate of the organophosphate pesticide methyl parathion in simple sediment–water microcosms (Pritchard *et al.,* 1983b) will help illustrate the virtue of this quantitative approach. In these experiments, methyl parathion (20 ppb) was continuously fed into glass vessels containing a 6-cm-deep plug (11-cm diameter) of saltmarsh sediment covered with 400 ml seawater; the distributions of the parent compound and of its major degradation product, *p*-nitrophenol, were followed with time. The water column in these systems was completely mixed but left the sediment–water interface intact. Other studies in our laboratory have shown that functionally, this microcosm is representative of a particular part of a local saltmarsh. The changes in concentrations of methyl parathion and *p*-nitrophenol in sterile (2% formalin) and nonsterile microcosms are shown in Fig. 7. Without any further information, these microcosm results mean very little environmentally; the results could not be used for quantitative predictions of how this saltmarsh could handle a continuous input of this pesticide. For example, hydrodynamic aspects of the saltmarsh have not been considered, nor has the relative importance of mitigating fate processes been determined. It is not known which conditions in the saltmarsh ecosystem will affect the rates of these processes.

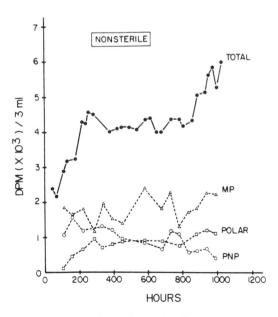

Figure 7. *(continued)*

Thus, two additional activities must accompany the microcosm study: a simple process model needs to be developed and some reductionistic analysis must be undertaken.

6.1. Model Development

Figure 8 shows a conceptual diagram of the potential pathways and distributions of methyl parathion and p-nitrophenol in the microcosm. It is a compilation of several assumptions about the fate of this chemical in natural systems. Each of the reactions (indicated by arrows) can ostensibly be described by a kinetic expression. A summary of the expressions used is shown in Table I.

The expression for biodegradation is particularly interesting, the major assumption being that the rate of biodegradation will be a second-order reaction expressed as some function of a rate constant, the chemical concentration, and the concentration of the catalytic unit. The catalytic unit is presumably

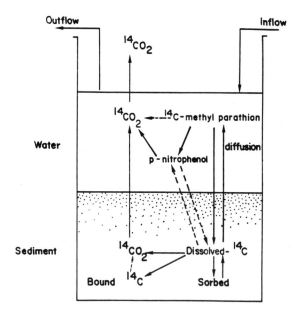

Figure 8. Conceptual description of mathematical model developed to describe the distribution of radioactivity, methyl parathion, and its degradation product, p-nitrophenol, in the microcosm. It requires kinetic inputs on hydrodynamics, diffusion into sediment, adsorption, desorption, partition coefficients, irreversible binding to sediment, abiotic degradation, biotic degradation, and biological mineralization. A mathematical equation for each of these processes has been formulated (see Table I).

Table I. Physical Parameters, Kinetic Expressions, and Rate Constants for Fate Processes Used to Describe the Distribution and Degradation of Methyl Parathion in a Continuous-Flow Microcosm

Physical parameter	Water column[a]	Sediment[b]			
		1	2	3	4
Depth (cm)	7.5	0.2	1.0	1.0	2.5
Volume (ml)	600	78.5[c]	78.5[c]	77.3[c]	16.9[c]
Diffusion coefficient (cm^2/sec^1)	C.M.	0.1	0.1	0.05	0.05
Detritus (g)	0	32.8	76.4	57.2	0
Sand (g)	0	0	0	240	721
Porosity (g/ml)	1.0	0.97	0.92	0.60	0.25
Hydroxyl concentration ($\times 10^7$)	3.5	0.63	0.63	0.62	0.63

Process	Kinetic expression[d]	Rate constant
Alkaline hydrolysis	$\dfrac{dc}{dt} = -K_{hyd}[OH^-][c]$	-2×10^4 M/hr
Biodegradation In water[e]	$\dfrac{dc}{dt} = -K_{Bio-w}[c]$	-0.0006 mg·hr/liter
In sediment	$\dfrac{dc}{dt} = -K_{Bio-s}[c][s]$	-0.004 liter/g per hr
Sorption rate[f]	$\dfrac{dc}{dt} = -K_{ADS}[s]$	Instantaneous
Equilibrium partitioning	$K_{oc} = \dfrac{[c]_{sorb}}{[c]_{diss}}[OM]^{-1}$	200 liters/kg

[a](C.M.) Completely mixed. [b]Segment 1 is the topmost segment. [c]Interstitial water.
[d][c] = chemical concentration; [s] = sediment concentration (mg/liter); K_{oc} = equilibrium partition coefficient normalized against organic matter associated with sediment; $[c]_{sorb}$ = equilibrium concentration of sorbate on sediment; $[c]_{diss}$ = equilibrium concentration of dissolved sorbate; [OM] = fractional mass of organic carbon in sediment.
[e]Biodegradation in water was treated as a first-order reaction, since the concentration of bacteria in microcosm and flask studies was equal.
[f]Adsorption and desorption rates are assumed to be completely reversible.

bacteria, but it could be other factors, such as free enzymes. Bacterial populations or communities will consist of active degraders, some of which will be more active than others. The degraders, under proper conditions, may also change in number due to growth or predation. Paris *et al.* (1981), on the other hand, have proposed that the biodegradation of pesticides in aquatic environments occurs under nongrowth conditions for the bacteria and that measurements of total colony-forming units on complex media will suffice for the concentration of the catalytic unit. (This assumes that the degraders are always a constant proportion of total population.) They have been able to attribute variations in biodegradation rate of three pesticides in 31 sampling sites to the variations in total colony-forming units. Their proposal is still relatively untested and probably compound-specific, but has encouraged research into the problem of measuring the catalytic units and developing kinetic expressions of biodegradation. Lewis and Holm (1981), on the other hand, have shown that under certain conditions, the biodegradation kinetics of methyl parathion cannot be described using measures of total microbial biomass. Bourquin et al. (1981) have reported similar complications. Spain et al. (1980) have shown that for p-nitrophenol degradation, a quantitation of the p-nitrophenol degraders is more appropriate than total biomass.

Once kinetic expressions have been formulated around the necessary assumptions, they can be incorporated into a process model, using the principles of conservation of mass of the general expression

$$V_1 \frac{dc_t}{dt} = Qc_{in} - Q_{out} + E'_{1,2}(f_{d_2}c_{t_2} - f_{d_1}c_{t_1}) - R_1$$

where V_1 equals the volume of the water column, c is the chemical concentration, Q is the flow through the water column, $E'_{1,2}$ is the exchange coefficient between the water column and the interstitial water of the top layer of sediment, f_d is the fraction of the total chemical that is dissolved [i.e., after adsorption to sediment, which is a function of the partition coefficient of the chemical and the organic matter content of the sediment (Karickhoff et al., 1979)], and R_1 equals the summed (assuming no interactions) kinetic expressions in the water (Pritchard et al., 1983b). Similar expressions account for movement of the chemical from water to the sediment and within defined sediment layers.

6.2. Reductionist Analysis

With a model in hand, reductionistic analysis is required; a fate study involving the determination of rate constants for each reaction must be undertaken. This is accomplished most simply by sampling water and sediment from the microcosm, placing it in a shake flask, and introducing methyl parathion. Concentrations of the parent compound and p-nitrophenol are followed with time. Through such studies, sorption rates, equilibrium partitioning, and biotic and abiotic degradation can be differentiated. The rate-limiting variables can also be examined. Abiotic degradation will be pH-dependent (Wolfe et al., 1977); partitioning to sediment will depend on organic content (Karickhoff et al., 1979). The relationship between biodegradation in the presence or absence of sediment can also be ascertained. Some of the rate constants derived from our flask studies are shown in Table I. A first-order rate constant was used for biodegradation in water; total biomass was not found to be significantly different between flask studies and microcosm water. A second-order rate constant was used for biodegradation in sediment; catalytic units were equated to sediment mass, assuming an equal distribution of bacteria on each sediment particle.

Figure 9 compares the model's predictions, using the kinetic information from the flask studies, with the actual microcosm data. In both sterile and nonsterile conditions, a relatively good fit of the model to the data was obtained ("good fit" is a nebulous term, since it is subjective at this point; differences of less than 4-fold are considered good). From this exercise, several conclusions can be drawn:

1. The second-order description of biodegradation in sediment using sediment mass as a quantitative measure of catalytic units proved sufficient to describe the microcosm data. Subsequent studies have shown that a biological "binding" process accounts for the degradation of methyl parathion (Pritchard *et al.*, 1983b) and probably some of the *p*-nitrophenol (unpublished data), thus corroborating the use of sediment mass as a surrogate for microbial biomass.

2. The kinetic expression for hydrolysis was best described as a second-order rate dependent on hydroxyl ion concentration. Fitting of the model to the data could not have been done without considering that pH in the microcosm

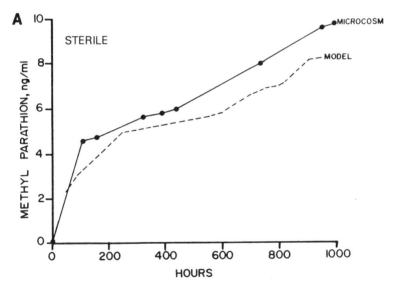

Figure 9. Comparison of changes in methyl parathion concentration in the water column (————) of the sterile (A) and nonsterile (B) microcosm with the predicted response of the microcosm model (------). Model prediction was based on rate coefficients derived from shake-flask studies (containing sediment–water slurries), on sorption rates and partitioning to sediments, and on biotic and abiotic degradation rates. An abiotic degradation rate constant of 2.0 $\times$ 10^4 M [OH]/hr and biotic degradation rates in water and sediment of 0.0006 mg·hr/liter and 0.002 liter/g per hr, respectively, were used in the model. A hypothetical biodegradation rate of 0.008 liter/g per hr was used in the second model line in the nonsterile microcosms to fit the data more closely. Methyl parathion concentrations were taken from Fig. 7. After Pritchard *et al.* (1983b).

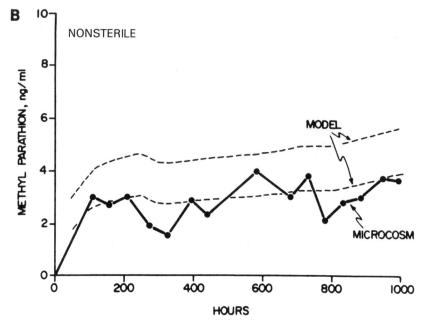

Figure 9. (*continued*)

in the sediment and at the sediment–water interface was 6.8 compared with 7.5 in the water column, thus necessitating two hydrolysis rates. Similar pH differences have also been recorded in the field.

3. Physical diffusion of the pesticide from the water compartment to the sediment compartment was correctly derived from the sterile microcosms. Diffusion into sediment is difficult to estimate, since water-mixing and turbulence above the sediment are the dominant forcing function for this process (Webb and Theodor, 1972). Mixing and turbulence in the field are probably more complicated than in the microcosm; thus, careful calibration of the microcosm with the field must be undertaken (see Section 9). Diffusion (molecular and mechanical) could therefore critically govern both fate and toxic exposure of a pollutant in nature.

4. Since the process model for the microcosm treated all fate processes additively, the fit of the model to the microcosm data suggests that this assumption was correct and that there were no subtle interactions or synergisms among fate processes that needed to be included in the model.

5. Finally, the microcosm verified to some extent the kinetic expressions used in the model and the environmental significance of the methods used to establish kinetic parameters. To make the predictive jump to the field, the microcosm study is set aside. An ecosystem simulation model (adjusted for the

conditions of a salt marsh), such as that proposed by Weigert *et al.* (1975), Clesceri *et al.* (1977), or Burns *et al.* (1982), can then be used to predict the fate of methyl parathion. This model will contain the same "verified" kinetic expressions as those discussed above and will take data from the "verified" flask studies, using natural samples from the saltmarsh to predict the fate of the pesticide in time and space from a known input into the saltmarsh. If the microcosm fulfills our expectations as the verification tool, then our conceptual basis for the extrapolation of laboratory data will be tenable. Our confidence will increase with experience and, where possible, with comparisons to actual field-dosing experiments.

6.3. Other Studies

Several other studies have used microcosms in a similar quantitative manner. Falco *et al.* (1977) were able to show that second-order kinetic expressions for biotic and abiotic degradation of malathion were sufficient to predict, through the use of a mathematical model, the fate of this pesticide in a flowing-stream microcosm. The microcosm was essentially an elaborate chemostat, having only a water compartment containing microbial and phytoplankton communities and receiving predetermined inlet concentrations of ammonia, nitrate, magnesium, and trace elements. Interestingly, total bacterial numbers were used in determining second-order rate constants in the flask, chemostat, and microcosm studies. The close agreement of the rate constants (within a factor of 4) suggests a relationship between bacterial numbers and the degradation rate that is independent of the population composition.

Marshall and Roberts (1971), using a process model to describe adsorption, biotic degradation, photolysis, and fish uptake for the organophosphate pesticide fenitrothion, were able to adequately simulate the change in concentration of the pesticide with time and its eventual distribution in a small farm pond. They concluded that laboratory estimates of the kinetic parameters affecting the fate of fenitrothion were an accurate reflection of the process dynamics in natural systems (the authors did not experimentally determine the kinetic parameters themselves, but rather retrieved the data from the literature). A microcosm study was not necessary in their case because residue analysis information from a field source that had received a dosage of the pesticides was also available from the literature.

The fate of the organophosphate pesticide chlorpyrifos (Dursban) in a pond has been assessed in a similar manner with equal success (Blau and Neely, 1975). Lassiter (1975) showed that simply the construction of a process model, without relating it to the field or to microcosm studies, is sufficient to point out the lack of good kinetic expressions for describing the microbiological and sorptive fate of mercury in aquatic systems. Wolfe *et al.* (1982) used the exposure analysis modeling system (EXAMS) model to predict the fate of hexachlorocyclopentadiene in hypothetical simulations of aquatic environments and showed that photolysis and, to a lesser extent, hydrolysis are the principal fate processes affecting the water and sediment concentrations of this chemical. It would be interesting to see whether their conclusions are correct. Since actual dosing of a field site with this toxic chemical would not be allowed, a sediment–water microcosm could be used as a verification of their predictions.

Continual demand for fate information on xenobiotics requires that this quantitative approach to microcosms be exercised frequently. Many other questions about the environmental conditions that affect biodegradation processes could be analyzed quantitatively in microcosms. Many pollutants, for example, are degraded through different pathways and at different rates, depending on the oxygen concentration or redox potential (Shaw and Hopke, 1975; Virtanen et al., 1982; Boyle et al., 1980; Liu et al., 1981; Yockim et al., 1980). Natural environments are typically characterized as having gradients of these factors, and a microcosm is the only means for establishing such conditions in the laboratory and examining their quantitative relationships to biodegradation for the purposes of predicting the fate of the chemical in the field. Biodegradation activities controlled by the exudates from plant root systems (Hsu and Bartha, 1979), by litter decomposition processes (Cummins, 1974), and by bioturbation (Hargrave, 1970) are also potential research areas in which the quantitative use of microcosms could be applied.

7. Quantitative Approach to Using Microcosms in Toxic-Effect Studies

Deleterious effects of pollutants on bacteria have been historically examined in relatively simple experimental systems, primarily shake-flask cultures using inocula that are unrepresentative of their original state. Measurements of pollutant effects on the microbial cycling of carbon, nitrogen, sulfur, and phosphorus are some of the most common assays performed in this way. The potential toxic effects of a pollutant on the nitrogen cycle can serve as a good illustration. Toxicologically, the nitrification part of this cycle is a critical effects end point, since the species diversity of nitrifying bacteria is quite narrow. If these organisms are killed, there is little functional redundancy in the system to accommodate the loss, and recovery may be very slow. This is generally an atypical situation relative to other microbially mediated cycles and heterotrophic processes.

Pure- and mixed-culture studies on the processes of nitrification, denitrification, and nitrogen fixation have been used numerous times as bioassays for the potential toxic effects of pollutants (e.g., see Atlas et al., 1978). Studies with pesticides (Tu and Miles, 1976; J. R. Anderson, 1978), herbicides (Greaves et al., 1976), heavy metals (Babich and Stotzky, 1980), and other pollutants (Colwell, 1978) have demonstrated some type of toxic effect on microbial populations, but by and large, the effects have been bacteriostatic and have occurred at environmentally unrealistic concentrations (10–500 mg/liter). Very few toxicity studies on the nitrogen cycle have dealt with the response of the microbial communities under conditions where the physical integrity of the natural system is preserved and the linkage with other trophic states is maintained.

7.1. Role of the Physical Integrity of Sediment in Nitrogen Cycling

Preserving the physical integrity of a natural environmental sample assures that the redox potential (Eh) gradients and dissolved oxygen profiles of sediments and soil are undisturbed. Nitrogen transformations are uniquely linked to the availability of electron acceptors, nitrification and denitrification being basically aerobic and anaerobic processes, respectively. Great confusion currently exists in the literature about exactly how the natural distribution of Eh in soil and sediment quantitatively relates to rates of nitrification and dentrification, since microniches apparently permit either process to operate irrespective of the surrounding oxygenation conditions (Painter, 1970; Focht and Verstraete, 1977). Thus, a definite need exists to implement microcosm studies with intact environmental samples so that the integrated effect of these microhabitats can be realized and studied relative to data obtained from conventional mixed-culture studies. From a toxicological point of view, searching for an inhibitory effect of a pollutant on an isolated component of the nitrogen cycle, i.e., using shake-flask cultures to measure nitrification, may represent only a part of the potential impact. Any interference with the biological control of Eh gradients and availability of electron acceptors in soils and sediments by a pollutant can greatly affect the dynamics of the nitrogen cycle. Henriksen *et al.* (1981), for example, have shown that many nitrifying bacteria in marine sediments, which are normally inactive due to oxygen (or Eh) limitation in a structured sediment situation, become very active in an aerated mixed-sediment procedure. They showed that correcting for reaeration in the mixed-sediment procedure (Henriksen *et al.*, 1981), or using intact sediment cores (Henriksen, 1980), produced nitrification rates similar to those observed in the field. Charyulu *et al.* (1980) showed that 2-aminobenzimidazole (a pesticide breakdown product) disrupted the Eh gradient in soil and thereby disturbed the nitrogen cycle through excessive stimulation of nitrogen fixation. Thus, indirect effects of a pollutant on the nitrogen cycle are possible, and most important, the effect would go unnoticed in conventional mixed-culture tests containing sediment slurries. These considerations also demonstrate the need to understand the kinetics of the nitrogen transformation processes *in nature* and to know their dependency on certain environmental variables. In the work of Henriksen *et al.* (1980), a quantitative relationship between oxygen concentration and nitrification rate was established and then essentially verified in a micro-cosm. Other processes in the nitrogen cycle could be treated similarly.

7.2. Role of Higher Trophic Levels in Nitrogen Cycling

Maintaining the linkage between nitrogen flux from sediment and the activities of various trophic levels in the water column is also an important kinetic equilibrium factor in the nitrogen cycle that could potentially be the focal point for disruption by a pollutant. This linkage is based on the supposition that in certain environments (such as coastal marine systems), organic nitrogen input into sediments, through excretions by zooplankton and decomposition of plant, animal, and bacterial remains, is balanced with inorganic nitrogen efflux from sediment and its eventual biological assimilation in the overlying water column. That is, the passage of a significant fraction of organic matter through benthic food chains and its eventual microbial remineralization in sediments controls primary productivity in coastal waters (Nixon, 1981). Seitzinger *et al.* (1980) have shown, using intact sediment–water cores, that Eh conditions permit significant amounts of inorganic nitrogen to be lost from the nitrogen cycle through denitrification in coastal marine sediments (nitrogen fixation is generally insignificant in these environments). Nutrient budget calculations indicate that this loss may account for the low fixed $N:P$ ratios typically observed in coastal marine ecosystems. Nixon (1981) believes that this low ratio is the basis for nitrogen-limited primary productivity in these areas. Thus, it is not unreasonable to imagine that any pollutant-related perturbation or disturbance to these physiochemical conditions under which nutrient cycling occurs will have a reverberating effect on ecosystem function and productivity.

Burrowing and bioturbation of sediments by various macrofaunal organisms are important in establishing the redox discontinuity layer (Edwards and Rolley, 1965; Fry, 1982) (i.e., the availability of oxygen, which would inhibit denitrification), as well as in mixing organic matter down into sediment where active denitrification is occurring (Hylleberg and Henriksen, 1980; McCall and Fisher, 1980). The balance between nitrification and denitrification is therefore probably tightly coupled to the sediment-reworking activities of benthic invertebrates (Chatarpaul *et al.*, 1979). Thus, effects of a pollutant on nitrogen cycling may be the result of an indirect response of the microbial populations to physical and biological conditions created by the activities or inactivities of other higher organisms.

A microcosm with an intact sediment core provides a good laboratory system in which natural sediment organization and oxidation–reduction profiles can be maintained for experimental purposes. This type of microcosm has been applied fruitfully to aquatic (Henriksen, 1980; Seitzinger *et al.*, 1980; Cappenberg, 1974; Pamatmat, 1971) and terrestrial (Van Voris *et al.*, 1980; Jackson *et al.*, 1979) nutrient cycling studies. It offers additional advantages in that the quantitative examination of the couple between water column primary productivity and remineralization processes in sediment can also be undertaken (Nixon et al., 1979). In addition, as stated earlier, the microcosm can be dosed with a toxicant in a manner that is simpler, more controllable, and environmentally safer than a similar dosing in the field.

7.3. Case Study: Effect of Oil Pollution on Nitrogen Cycling

A study to exemplify how a microcosm may be used to reveal a unique toxic response of a microbial community is found in the work of the University of Rhode Island's MERL research group, which deals with the effects of crude oil on a coastal marine ecosystem (Elmgren et al., 1980). The researchers successfully modeled Narragansett Bay (Rhode Island) in very large outdoor tanks (13 m^3) (Pilson et al., 1979, 1980). Tanks (5.5 m high, 1.83 m in diameter) contained a box core of Narragansett Bay sediment (30 cm deep) that received continuous unfiltered bay water (turnover time 27 days) while exposed to natural light and temperature regimes. Stirring devices were designed to direct turbulent energy onto the sediment to effect resuspension of flocculent material as observed in the bay. The size of these tank systems, although originally designed to satisfy a specific sampling regime, greatly restricts experimental manipulation. Further, their cost and technical support requirements cannot be accommodated by most laboratory research programs. However, these systems represent some of the most extensively characterized long-term enclosure-type experiments to date, providing detailed descriptions of annual plankton and nutrient cycles, benthic invertebrates, turbulent mixing, and replicability (see Giesy, 1980). They are among the few complex systems to be dosed with a toxicant (Elmgren et al., 1980). For the sake of our illustration, assume that similar effects on the nitrogen cycle can be demonstrated in small laboratory microcosms. This is probably a valid assumption because Adler et al. (1980) and Nixon et al. (1979) have indicated that much of the same ecosystem structure and function in the MERL tanks (coupling between pelagic and benthic components) can be modeled in 150-liter microcosms.

The toxicant study consisted of a chronic (5-month continuous), low-level (range 60–350 ppb) dosing with No. 2 fuel oil. Significant effects were observed in both the planktonic and benthic communities (Elmgren et al., 1980). Phytoplankton populations showed a significant increase in biomass and a radically different species composition relative to control tanks. The observed changes were probably a result of decreased grazing pressure caused by a toxic effect on the zooplankton and benthic suspension feeders. Likewise, in the sediment, benthic protozoan populations increased dramatically as a result of substantial decreases in predatory macrofaunal populations. In addition, bioturbation activities in the sediment were significantly reduced due to the drop in macrofaunal numbers. Most important from the standpoint of microbial activity was a major suppression in the production of total inorganic nitrogen (NH_4^+, NO_2^-, NO_3^-) from the sediments. This is shown in Fig. 10. Similar reductions in nitrate generation from beach sand microcosms treated with oil have also been reported (Harty and McLachlan, 1982). The importance of this effect is apparent, considering that the principal growth-limiting factor for phytoplankton productivity in Narrangansett Bay is controlled by the rate of mineralization of organic nitrogen deposited from the water column into the sediment and the eventual efflux of nitrogen from sediment (Nixon, 1981).

Elmgren *et al.* (1980) demonstrated that effects observed in their study could not have been predicted from knowledge of the biology and susceptibility to oil of the individual species. However, further dissection and analysis of the test results are required to provide greater comprehension of the interactions

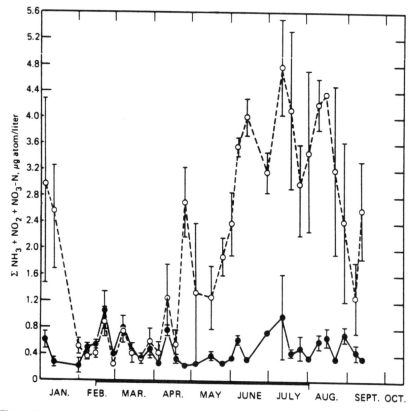

Figure 10. Average concentration of total inorganic nitrogen per liter ($\pm$S.E.) in three oiled tanks (●—●) and three control tanks (○---○). The heavy line on the time axis denotes the period of oil addition. After Elmgren *et al.* (1980).

involved and to develop a predictive model to forecast similar toxic effects in other environments.

7.4. Model Development

Quantitative use of a microcosm to study the toxic effects of a pollutant can be performed, in theory, by first establishing a state-of-the-whole in the microcosm and then measuring the effect of a pollutant on that state. Extrapolation of this potential effect to the field necessitates a series of research steps similar to those described for fate studies.

A conceptual process model for the microcosm results must be developed to clarify the kinetic relationships among system components. The model should reflect current knowledge and assumptions about the processes suspected to operate in the system of interest, particularly those that are possibly sensitive to the pollutant. All the kinetic expressions for the system processes, pathways, and component interactions necessary to provide the required outcome must be represented.

Rate information for each process involved must be obtained by using either values from the literature or experimentally determined values. Incorporation of this information into the process model will yield predictions (such as nutrient flux from sediment) that can be compared with the undosed microcosms. If the model is unable to predict the microcosm results within some degree of error, then the iterative process outlined in Fig. 2 can be used to modify the kinetic expressions for the model.

Once the model has been adjusted to provide reasonable predictions, various rate functions can be adjusted to simulate positive and negative responses to a pollutant input, and the end result can be compared with the results from the dosed microcosm. The implicated processes can then be experimentally tested, assuming that the laboratory methods for their isolation and study have been worked out, for their sensitivity to the pollutant. Dose–response relationships can subsequently be established.

In the oil-dosing experiment described in Section 7.3, it is reasonable to hypothesize that the reduction in total nitrate efflux from sediment could have been caused by a change in the redox profile of the sediment resulting from a toxic effect of the oil on bioturbating macrofauna. Bacterial populations themselves are generally not chronically poisoned by oil pollutants (Colwell, 1978). A change in the redox profile, however, could dramatically increase dentrification rates, since it is already believed that 5–20% of the nitrogen remineralized in these sediments is lost as N_2 and N_2O (Seitzinger et al., 1980). Studies have shown that pollutants will affect the bioturbation activities of benthic invertebrates. Rubinstein (1979), for example, has shown that Kepone at sublethal concentrations inhibits sediment-reworking activities of the lugworm (Arenicola cristata); this was observed in a bioassay system in which the extent

of detrital loss from the surface of sand (i.e., incorporated into the animal burrow) was measured photographically. A linear dose dependency was also
observed.

The hypothesis concerning increased denitrification due to oil pollution
could be tested with a process model describing the relationship among biological sediment reworking, control of Eh/dissolved oxygen profiles, and control
of nitrogen remineralization. To our knowledge, this has not been attempted,
although available literature seems to indicate that the task is feasible. A possible, highly simplified conceptual model of this role of bioturbation in nitrogen
cycling is shown in Fig. 11. Included is a model of the nitrogen cycle developed
by Lassiter (1975) and a two-layer model distinctly dividing the nitrification
zone from the denitrification zone (Billen, 1982; Vanderborght and Billen,
1975) and a bioturbation model reported by Hylleberg and Henriksen (1980).
Each arrow represents a process or group of processes for which some kinetic
expression and some method of rate measurement are needed. Most of these
expressions take the form of biomass-dependent second-order reactions:

$$k_2 = [b]e^{-k_1 t}$$

where k_2 is the apparent rate constant, k_1 is the exponential substrate dependence term, and b is the concentration of biomass, for example, bacterial or
invertebrate population size. However, a number of mechanistic models
describing nitrogen transformations in soil have assumed first-order (with
respect to substrate concentrations) reactions (Mehran and Tanji, 1974). Others have factored in microbial growth using the Monod equation. Bazin et al.
(1976) have presented an excellent review of those nitrogen transformation
models developed for soil systems. In Lassiter's model, oxygen concentrations
are used to switch the nitrification and denitrification processes on and off, simulating aerobic and anaerobic conditions. Each major dynamic variable (see
Fig. 11) can be expressed as a sum of all the controlling parameters. For example, the parameters used to describe the changes in concentrations of ammonium, nitrite, and nitrate, and their distribution relative to the two-layer model
of Vanderborght and Billen (1975), are shown in Tables II and III.

Values for each parameter can be assigned either by application of literature values or by appropriate experimentation. Since the literature on the
microbiology of the nitrogen cycle is extensive (Focht and Verstraete, 1977),
much of the process information can come from that source. Hopkinson and
Day (1977) have listed many of the rates typical for nitrogen cycle processes
in coastal marine environments. In other cases, experimentation will be
required. The work of Henriksen et al. (1981) that showed the relationship of
oxygen concentration to the nitrification rate has been cited. Jahnke et al.
(1982) successfully modeled the quantitative dependence of denitrification on
sediment respiration, using profiles of pore-water nutrients. Hansen et al.
(1981) and Graetz et al. (1973) have demonstrated the relationship of nitrifi-

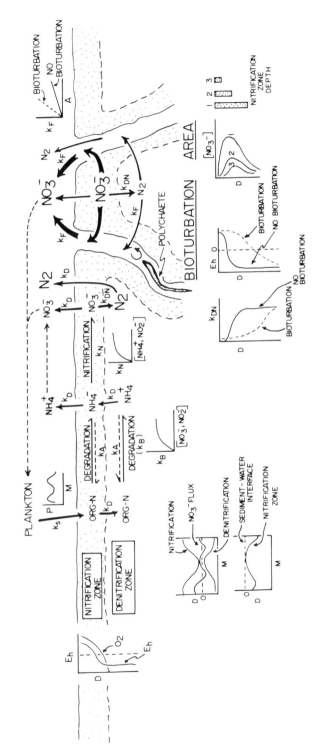

Figure 11. Conceptual framework of nitrogen cycling in coastal marine sediments based on the two-layer model of Vanderborght and Billen (1975) as affected by bioturbation (Hylleberg and Henriksen, 1980). (A) Area; (D) depth; (k_A) assimilation rate; (k_B) biodegradation rate; (k_D) diffusion rate; (k_{DN}) denitrification rate; (k_F) flux rate due to bioturbation and diffusion; (k_N) nitrification rate; (k_S) settling rate; (M) months; (ORG-N) organic nitrogen; (P) production.

Table II. Summary of the Kinetic Expressions Used to Describe the Cycling of Ammonium, Nitrite, and Nitrate in Natural Systems and to Exemplify the Tight Coupling of Associated Processes with Oxygen Availability[a]

$$\frac{d[NH_4^+]}{dt} = Sk_i I[O_2][P_i][ORG\text{-}N][NO_3^- \text{ or } NO_2^-]$$ 　(1) Anaerobic decomposition of organic nitrogen using nitrate reduction (denitrification)

$$+ k_i[P_i][ORG\text{-}N][O_2]$$ 　(2) Aerobic decomposition of organic nitrogen

$$- k_i[P_i][NH_4^+ \text{ or } NO_2^-]$$ 　(3) Nitrogen fixation

$$- k_i[P_i][NH_4^+ \text{ or } NO_2^-][O_2]$$ 　(4) Production of organic nitrogen through assimilation of ammonia or nitrite

$$- k_i I[O_2][P_i][ORG\text{-}N][NO_3^- \text{ or } NO_2^-]$$ 　(5) Assimilation of nitrate during anaerobic decomposition of organic nitrogen

$$- k_i[P_i][NH_4^+][O_2]$$ 　(6) Direct nitrification of ammonia to nitrite

$$\frac{d[NO_2^-]}{dt} = k_i[P_i][NH_4^+][O_2]$$ 　(7) Input by nitrification

$$+ Sk_i I[O_2][P_i][ORG\text{-}N][NO_3^-]$$ 　(8) Input by denitrification
$$- Sk_i[P_i][NO_2^-][O_2]$$ 　(9) Loss via nitrification
$$- Sk_i I[O_2][P_i][ORG\text{-}N][NO_2^-]$$ 　(10) Loss via denitrification

$$\frac{d[NO_3^-]}{dt} = k_i[P_i][NO_2^-][O_2]$$ 　(11) Input by nitrification

$$- (Sk_i I[O_2][P_i][ORG\text{-}N]$$ 　(12) Loss via denitrification
$$+ k_i I[NH_4^+][P_i][O_2][NH_3^- \text{ or } NO_3^-][NO_3^-]$$ 　(13) Assimilation by various microbial populations

[a]After Lassiter (1975).

Assumptions:

a. The synthesis of organic nitrogen material (ORG-N) uses nitrogen obtained from four sources: ammonia, nitrite, nitrate, and N_2 fixation. Organic nitrogen was a fixed portion of total organic matter: 10 moles of N atoms per mole of organic material. Appearance and disappearance of organic nitrogen was dynamically expressed as

$$\frac{d[ORG\text{-}N]}{dt} = \Sigma k_d[P_i] - \Sigma k_b[P_i]I[O_2][ORG\text{-}N]$$

where K_d is the death rate of any microbial population $[P_i]$ and k_b is the decomposition (aerobic and anaerobic) rate of microbial population $[P]$ as a function of oxygen availability ($I[O_2]$).
b. Ammonia is used as a suppressant for N_2 fixation [equation (3)].
c. Sizes and chemical composition of microbial populations (P_i) are fixed. The nitrogen-fixing population is active only when nitrate and ammonium are scarce.
d. Oxygen is a forcing (time-varying) function causing reactions to be turned on or off. Oxygen inhibition functions (I)—which could mean, for example, actual inhibition, such as inhibition of denitrification by the presence of oxygen, or a preference, such as inhibition of NO_3^- uptake by microorganisms in the presence of "preferred" NH_4^+ were expressed kinetically as

$$k = k_{max}e^{-ax^2}$$

where k is the rate coefficient of the inhibited process, x is the concentration of inhibiting substance, and a is a scaling coefficient.
e. S is a stoichiometric coefficient and k_i's are rate coefficients for various processes.
f. Assimilation of nitrate, ammonia, and nitrite accompanies microbial processes in which biomass generation is energetically feasible (aerobic decomposition, denitrification, and nitrification). Assimilation terms for both ammonia and nitrate are represented as proportional to the product of the energy-yielding process and the concentration of NH_4^+.

Table III. Summary of the Kinetic Expressions Used to Describe Nitrate Concentration in Intersititial Water between the Nitrification and Denitrification Zones of Muddy (Organic-Rich) Sediments[a]

Stationary profile model

$$[NO_3^-] = [NO_3^-]_0 e^{-(k_d/D)^{1/2}z}$$ (1) Nitrate distribution profile in sediment

Assumptions:

a. $[NO_3^-]$ is the most important limiting factor affecting the rate of denitrification, and denitrification can therefore be represented as a first-order reaction (k_d) with respect to $[NO_3^-]$ in pore water.

b. Nitrates diffuse in pore water according to Fick's first law, with a constant diffusion rate (D) and as a function of hydrodynamic turbulence and biological perturbations ($D = 10^{-5}$ to 5×10^{-5} cm^2/sec).

c. The nitrate concentration gradient at the sediment–water surface is negative; nitrates diffuse from overlying water into sediment.

Stationary two-layer model

$$[NO_3^-] = -(k_n/2D)Z^2 + A_z + [NO_3^-]_0$$ (2) Diagenetic equation in the nitrification zone

$$[NO_3^-] = Be^{-(k_d/D)^{1/2}(Z-Z_n)}$$ (3) Diagenetic equation in the denitrification zone (below Z_n)

$$A = \dfrac{\dfrac{k_n}{D}\left[\dfrac{Z_n^2}{2} + \left(\dfrac{D}{k_d}\right)^{1/2}Z_n\right] - [NO_3^-]_0}{Z_n + \dfrac{D}{k_d}}$$

$$B = (k_n/2D)Z_n^2 + AZ_n + [NO_3^-]_0$$

d. There is a limit level, Z_n, corresponding to the depth where Eh falls below the value for which oxidation of ammonium [nitrification (k_n)] is no longer energetic. Denitrification (k_d) will occur below this depth.

[a]After Vanderborght and Billen (1975).

cation rates to sediment depth, Eh, and season. Fry (1982) has discussed the methods for relating bioturbation to reaeration of sediments. Hylleberg and Henriksen (1980) and Henriksen et al. (1980) have quantitated the effects of bioturbation on nitrification rates. J. G. Jones (1979) has demonstrated a quantitative relationship between Eh values and denitrification rates (measured as nitrate reductase). Extensive measurements of redox potential in the sediments of the microcosms must also be carried out both with and without the oil exposure. This can be accomplished either with the dehydrogenase assay technique (Pamatmat and Bhagwat, 1973; Vosjan and Olanczuk-Neyman, 1977) or directly with platinum electrodes (Howes et al., 1981).

Finally, information will also be needed on measurements of sediment-reworking activities. They can be estimated indirectly from measurements of respiration by invertebrates in sediment (McCall and Fisher, 1980) as long as data are available on the density, physical movement, and burrow configuration of the principal bioturbating invertebrate. This information is available in the literature or can be obtained experimentally, using existing measurement methods.

With this information, the validity of the model can be iteratively tested by referencing the model outputs with results from the microcosm and readjusting the model (or the kinetic descriptions of the processes) until adequate predictions of the microcosm results are obtained (see Fig. 2). Once confidence in the predictiveness of the process model is acquired, the model can be integrated with an ecosystem simulation process model to account for spatial and temporal parameters (geomorphology, water hydrodynamics, reaeration by turbulent flow, deposition rates of organic nitrogen, variations in Eh gradients from site to site, steady state, and time-variable concentrations of inorganic nitrogen in the water column), and couplings between nutrient efflux from sediment with phytoplankton and zooplankton population dynamics (Nixon and Kremer, 1977), which cannot be observed or accomodated in small microcosms. The possibility then exists of demonstrating that the impact of small changes in the activity of a microbial community in response to a pollutant (depressed nitrogen efflux from sediment) can be magnified significantly after being coupled to other trophic activities of the ecosystem (primary productivity). Examples of ecosystem simulation models have been described by Clesceri *et al.* (1977), Hopkinson and Day (1977), and Nixon and Kremer (1977). It should be apparent that the rate measurements of the processes in the nitrogen cycle and a quantitative knowledge of the rate-controlling parameters, including the effect of a pollutant, are the results that are extrapolated to the field, not the microcosm results.

The importance of using models as interpretative tools and of using microcosms as representations of the state-of-the-whole in the study of activities of microbial communities cannot be overemphasized. Unfortunately, very few process modeling efforts have been coupled with a microcosm study in the area of microbial nitrogen cycling, or any other cycles for that matter. Even in the application of models to nitrogen transformations in soil, Bazin *et al.* (1976) have pointed out the need to produce more complete laboratory simulations of nitrogen turnover in natural systems so that more of the "black box elements" can be replaced with analytical descriptions of relevant processes.

7.5. Additional Examples

Model development may suggest a need to experimentally investigate a new process or interrelationship. For example, in working with nitrogen cycling in estuarine salt marshes, nitrogen fixation should probably be considered as a major factor in regulating the nitrogen budget. Mann (1979) showed that measurements of N_2 fixation in bell jars placed over *Spartina* and other saltmarsh plants in the field yielded rates considerably faster than those projected from a separate study of cyanobacteria taken from the mud surface. The bell jar experiments in this case provided the integrated ecosystem response, since the field site was physically accessible. If, however, the effects of a pollutant are to be investigated, a microcosm may be required (it is feasible to take intact sediment cores with associated macrophytes for this purpose). Laboratory studies by Mann (1979) indicated that N_2-fixing bacteria associated with the isolated root parts of *Spartina* (two bacterial types have been isolated, a facultative anaerobe and a microaerophile) were quite sensitive to oxygen concentrations (5–9% was optimal) and to the supply of dissolved nitrogen compounds in the surrounding milieu (Fig. 12). This explained the observed higher rate of N_2 fixation associated with the plants along the edges of the marsh that were exposed through tidal action to seawater containing dissolved nitrogen compounds.

From a process modeling standpoint, kinetic rate expressions could be developed for all sources of N_2 fixation and for the environmental factors controlling each source and then integrated into a model. The model can be checked to verify whether it predicts the N_2-fixation rates observed *in situ* or in a microcosm. By examining microcosm (or field) results and considering the regulation of microbial activity in relation to the physical integrity of natural systems and the interaction with higher trophic forms, a new site for the pos-

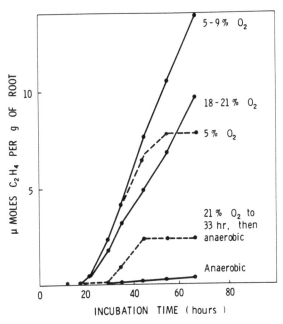

Figure 12. Nitrogen fixation (acetylene reduction method) by washed, excised roots of *Spartina alterniflora* at various oxygen concentrations. After Mann (1979).

sible adverse effects of a toxicant can be realized. It may be considerably more sensitive than the direct tests on pure cultures of cyanobacteria.

Another excellent example of applying a process model to a set of microcosm experiments is the work of McIntire on the dynamics of lotic periphyton in laboratory streams (McIntire, 1964, 1965, 1966, 1978, McIntire et al., 1975). His first approach was to operate the stream microcosms in the laboratory and thoroughly examine the periphyton community structure and population dynamics as a function of light intensity and current velocity. This was basically an exercise in manipulating an ecosystem to change its overall response (see Section 10). He then attempted to document the major controlling variables affecting the processes of primary production, community respiration, introduction of organic and inorganic materials from outside the system, export of organic and inorganic materials, and the grazing activities of aquatic animals. A kinetic expression was developed for each process based on the literature and the data from a variety of field and laboratory studies (McIntire et al., 1975; McIntire, 1978). Data describing the variables of the system, such as temperature, light intensity, current velocity, nutrient concentration, silt loading, and organism biomass, were incorporated into these expressions where appropriate. By then combining these mathematical descriptions into a process model and running the model under steady-state conditions, predictions of periphyton dynamics were compared with microcosm results as a means of validating the model.

Toxicological studies can now be applied to this analytical framework. For example, the effect of a pollutant on a microbial community responsible for the mineralization of organically bound nitrogen or phosphorus from leaf litter can be translated to an effect on periphyton dynamics in stream ecosystems. Similarly, the toxicity of a pollutant to grazers (snails, in this case) can be examined for "domino" effects on microbial mineralization rates and ultimately periphyton dynamics. Interpretation and field extrapolation of the results from stream microcosms exposed to a pesticide (S. R. Hansen and Garton, 1982) and to fuel oil (Bott and Rogenmuser, 1978) could probably be enhanced by applying the McIntire model.

Several other experimental situations have been reported that might allow a quantitative demonstration of the toxic effects of a pollutant on a microbial population. Fry and Ramsey (1977) reported on the indirect changes in heterotrophic activity of microbial epiphytes on two aquatic macrophytes, *Elodea canadensis* and *Chara vulgaris*, following treatment of the plants with the herbicide paraquat. The kinetic parameter $K + Sn$, where K is the half-saturation constant and Sn is the natural substrate concentration, for glucose mineralization was normally quite high (36–150 μg/liter) for epiphytes on *Elodea*, suggesting that healthy plants secrete dissolved materials that are then available to the bacteria. Killing the *Elodea* with paraquat significantly increased the $K + Sn$ and V_{max} (maximum glucose-uptake rate) of the epiphytes, which was not the case with the epiphytes on the paraquat-resistant *Chara*. The increase in kinetic parameters was not the result of increases in population size; rather, the physiological activity of each cell apparently changed. This coupling of microbial community responses with higher trophic level activities is another example of potential indirect effects of a toxicant on bacteria. It is a relationship that requires a natural physical association that could be accommodated in a microcosm.

The availability of methyl mercury in sediment is controlled by several processes operating simultaneously at a rate dependent on environmental parameters closely tied to physical structure of sediment beds and to physical interactions with other trophic activities. Since mercury cannot be used in the field, microcosm studies are required. Bacterial populations in sediment can convert elemental mercury to the mercuric ion form (Nelson and Colwell, 1975) and can then convert this product to methyl mercury (Jensen and Jernelov, 1969). The methyl mercury can subsequently be degraded by other bacterial populations (Billen *et al.*, 1974) that appear to adaptively increase their mineralization capacity with exposure to increasingly larger concentrations of methyl mercury (Billen *et al.*, 1974). This set of processes is affected by anaerobic conditions, by the amount of organic matter (Olsen and Cooper, 1976), perhaps by salinity (Blum and Bartha, 1980), and by the presence of invertebrates that can rework the sediment (Jernelov, 1978).

Without a process model and some complex natural system to use as an integrating tool, it is very difficult to make quantitative predictions regarding the net rate of methyl mercury accumulation in an ecosystem. Some effort has been made to develop kinetic expressions for each of the processes affecting the fate of elemental mercury in natural systems (Bisogni and Lawrence, 1975), and results from a microcosm study have been reported (Titus *et al.*, 1980). The latter study provides interesting information on distribution and partitioning of mercury between sediment and water. Without the use of a process model, however, the results from the microcosm have little predictive value.

8. Interaction between Microcosms and Simpler Systems

The holistic approach to studying a microbial community in the laboratory, when combined with the classic reductionist approach of studying small pieces of the whole (pure- or mixed-culture studies, for example), is a potentially productive way to increase understanding of microbial ecology, particularly as it relates to pollutant-mediated responses. Part of the beauty of this combination is that one approach can be graded into the other through a series of scaled successions, theoretically by either a "scale-up" or a "scale-down" from a complex natural situation.

In the "scale-up," specific types of pure and mixed cultures can be studied together in various combinations and the end result compared with the complex situation (state-of-the-whole). The intent is not to reproduce the whole exactly, but instead to compare the relative contribution of each pure- or mixed-culture component. This has been tried several times in microbial ecology with moderate success, although in many cases a state-of-the-whole or microcosm study was not included. A good example is the work of Coleman and colleagues (Coleman *et al.*, 1978a,b; Herzberg *et al.*, 1978; Anderson *et al.*, 1978; C. V. Cole *et al.*, 1978), in which a series of factorially designed experiments were employed to investigate the effects of trophic interactions on the mineralization of organically immobilized carbon, nitrogen, and phosphorus in soil. The trophic interactions were modeled as a simple food web consisting of various combinations of plant exudate and pure cultures of bacteria, amebae, and nematodes inoculated into sterilized soil. The researchers established that once phosphorus was immobilized by bacteria, it was rapidly remineralized by the saprophagic grazing of the amebae and subsequently released as inorganic phosphorus. Although these results are informative, their environmental validity could be greatly enhanced by a comparison with the complex state of a microcosm. The microcosm could be an intact soil core in which the leaching of inorganic phosphorus after exposure of the core to plant exudate could be measured. Since the activity per organism had been established separately, the biomass of the bacteria, amebae, and nematodes in the core could be determined and monitored with time. A simple model could then be used to see how well the individual pieces predicted the whole. The suitability of this particular example in measuring response to the addition of pollutants is obvious. A classic study of leaf litter decomposition and mineral cycling in forest ecosystems was carried out by Witkamp and colleagues (Patten and Witkamp, 1967; Witkamp, 1976; Witkamp and Ausmus, 1975), using the same type of factorial experimentation with a variety of simple components (leaf litter, microflora, millipedes, snails, and soil). Again, the synthesis of various degrees of complexity was never compared with the natural complexity in the field or in a microcosm.

The "scale-down" situation involves both manipulation of a complex ecosystem in the laboratory and experimentation with isolated microbial communities or subsystems that are derived by disassembling the complex system. This is likely to be a more productive situation that the "scale-up" exercise. The usefulness of the microcosm, in this case, is to isolate and illustrate important mechanisms that control the real system. Our studies on the fate of *p*-cresol illustrate how the complex results from a microcosm can be manipulated and disassembled into subsystems to provide quantitative information on the factors controlling the fate of a chemical in an aquatic system and provide input data for a mathematical model. In conjunction with the dosing of several man-made stream channels with the toxicant *p*-cresol (Cooper *et al.*, 1982), we performed laboratory studies on the fate of *p*-cresol to see how accurately we could predict its fate in the field streams. A microcosm study, involving the fate of *p*-cresol, was therefore conducted. The microcosm design (Fig. 13) used a tray technique (36 cm diam., 10 cm high) to transport intact portions of the riffle area and pool area in the streams to the microcosms. Plants associated with these areas were included in the trays. A recirculation device provided complete mixing of the water column (80 liters) and produced turbulence similar to that in the stream. The stream channel was actually a closed recirculating loop; consequently, the microcosms, as a riffle–pool pair, were maintained as closed systems. Since the stream was autotrophically based, diurnal dissolved oxygen profiles were used to calibrate the microcosm with the field. Invertebrate populations on the rocks and sediment were also used as calibration points. In both cases, a good simulation of the stream was obtained by the microcosms (Pritchard and Cripe, 1983).

The microcosms (riffle–pool) were spiked with 8 ppm *p*-cresol; the disappearance of the parent compound was followed by high-pressure liquid chromatographic analysis. The result (Fig. 14) is representative of the integrated

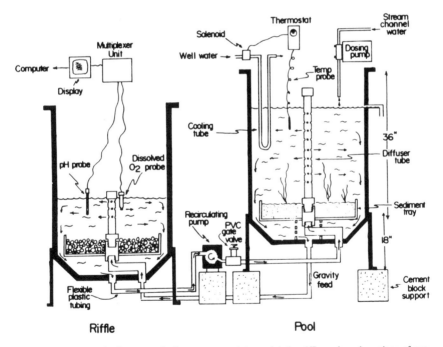

Figure 13. Schematic diagram of microcosms used to model the riffle and pool sections of outdoor stream channels. Each tank holds 80 liters of stream water. Microcosms are operated as closed systems in which recirculation and turbulence generation are accomplished through the use of a diffuser tube and a water pump. Trays are filled in the field and allowed to acclimate there for 2 weeks prior to their transfer to the microcosms. After Pritchard and Cripe (1983).

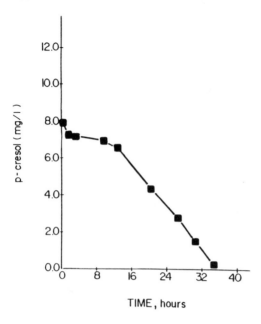

Figure 14. Rate of disappearance of p-cresol (initial concentration, 8 mg/ liter) in a riffle–pool microcosm. The microcosm was operated as a closed, continuously mixed reactor. After Pritchard *et al.* (1983a).

response of the system as a whole; the rate of disappearance of p-cresol is controlled by an unknown number of factors the integration of which produces the microcosm result. No loss of p-cresol occurred in azoic microcosms.

It is important to point out that this microcosm result has little predictive value of its own. The microcosm is not a duplicate of the stream, and the results have no quantitative bearing on the rate of disappearance of p-cresol in the stream channel. Therefore, only a qualitative judgment about its fate in the stream can be made. However, a quantitative extrapolation of the microcosm result to the field is possible if the controlling variables that affect the fate of the compound are demonstrated, i.e., kinetic information about the processes involved, specifically rate constants.

To understand more clearly the factors controlling the fate of p-cresol, the rate of degradation of p-cresol in microcosms containing only riffles and only pools was compared. It can be seen in Fig. 15 that the all-pool microcosms degraded p-cresol slower than the all-riffle microcosms. The degradation rate in the pool–riffle microcosm was intermediate. The manipulation indicated that stream riffle areas were quite important in the initial biodegradation of p-cresol. Respiking of the microcosms produced similar degradation rates in all systems (Fig. 15), indicating that the biodegradation contribution from the riffles is important only in initiating degradation of p-cresol.

Disassembling the integrated responses of the whole into subsystems (that

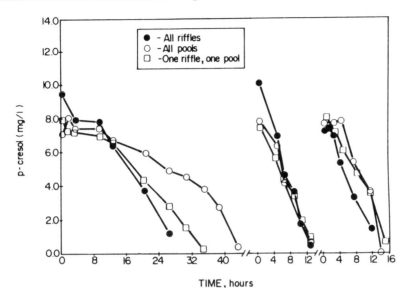

TIME, hours

Figure 15. Rate of disappearance of *p*-cresol (initial concentration, 8 mg/liter) from three types of microcosm sets: two riffle tanks (●), two pool tanks (○), and one riffle tank/one pool tank (□). At hours 45 and 60, all three microcosms were respiked with 8 mg/liter *p*-cresol. After Pritchard *et al.* (1983a).

is, microbial communities) to obtain more quantitative information was straightforward. Mixed microbial populations associated with water, sediment, rock surfaces, and plant surfaces were separated from the microcosm and studied in shake flasks to determine their *p*-cresol biodegradation potential (Pritchard *et al.*, 1983a). Figure 16 shows the degradation rates of *p*-cresol in shake flasks containing components from the stream. Bacteria associated with rock surfaces were the most active degraders. These subsystem or mixed-culture experiments provided degradation rates per unit mass or rate constants, i.e., rates per liter of water, rates per rock, or rates per gram of sediment. By knowing the approximate mass of each component in the stream channel, a prediction of the rate of disappearance of *p*-cresol during dosing can be made using a mathematical model [e.g., the EXAMS model (Burns *et al.*, 1982)]. Criteria for integrating our rate information came from the microcosms. Adaptation to faster degradation rates appeared to have been coupled to some factor associated with rocks or sediment, since adaptation in water-only shake flasks took much longer than in the microcosms (data not shown).

Another interesting form of the "scale-down" considerations is suggested in the adaptation studies of Spain *et al.* (1980). Adaptation of microbial pop-

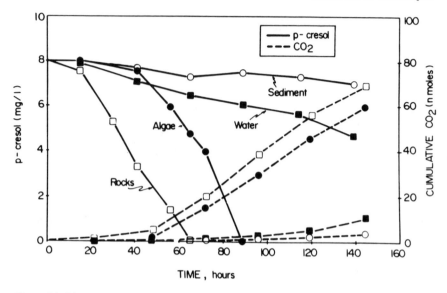

Figure 16. Disappearance (—) and mineralization (----) rates of p-cresol in shake-flask systems containing different components from the stream channels. For algae (*Spirogyra,* 162 mg wet wt./liter), sediment (340 mg wet wt./liter), and rocks (seven, 1.4 mg dry wt. of periphyton/ rock), components were washed three times with sterile stream water and then incubated in sterile stream water supplemented with p-cresol. After Pritchard and Van Veld (1983).

ulations to greater biodegradation activity of p-nitrophenol (PNP) was tested in a small microcosm using a sediment–water core from fresh and saltwater environments. The microcosm design is shown in Fig. 17. Estuarine environments in most cases have shown no adaptation within the same time frame needed to see adaptation in freshwater environments. In an attempt to better understand this lack of adaptation in estuarine environments, several potential manipulations could be undertaken. For example, Spain (personal communication) has isolated a PNP degrader from an estuarine site, using a selective enrichment technique. This presents the possibility of seeding an estuarine microcosm with the isolate to see whether adaptation can be provoked. If adaptation occurs, it suggests that environmental conditions in the microcosm do not preclude this event; rather, it would appear to be a deficiency in the indigenous microbiota (very low numbers of PNP degraders that might grow very slowly). If adaptation did not occur, it would suggest that the environmental conditions of an estuarine environment, relative to its freshwater counterpart, were not conducive to PNP degradation. Either way, this manipulation of a microcosm specifically directs the research toward determining why estuarine environments cannot degrade this pollutant.

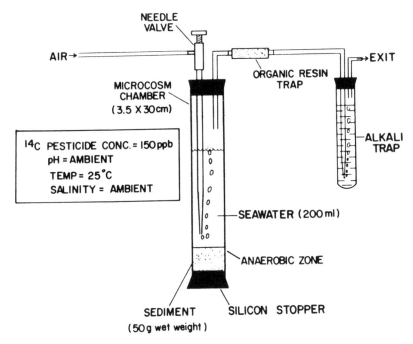

Figure 17. Schematic diagram of eco-core biodegradation test system used for biodegradation studies in estuarine saltmarsh environments. After Pritchard *et al.* (1979).

9. Field Calibration of Microcosms

Simulation of ecosystem components in microcosms requires some verification, generally by making comparisons to the field. Since exact simulation is unlikely, it is important to check various parameters and behaviors of a microcosm to determine whether they agree, within certain acceptable limits, with similar parameters and behaviors measured in the field. If the behavior of the microcosm falls outside these limits, then it is theoretically possible to change the microcosm design or operating conditions to bring the behavior within the acceptable limits. This exercise was referred to as the field calibration of microcosms in our studies (Pritchard, 1981). "Acceptable limits" are subjective evaluations related to the investigator's previous experience and to consensus judgments based on the literature. A calibrated microcosm should infer that a degree of similarity (or dissimilarity) between the laboratory and field has been established and can be factored into or integrated with the interpretation of microcosm results. Processes and behaviors in a microcosm may differ considerably in magnitude, but should show the same trends as the field.

9.1. Functional Attributes

There have been some efforts to establish the parameters and behaviors of an ecosystem that are most suitable as indicators of differences between microcosm and field (Perez *et al;* 1977; Harris, 1980; Jassby *et al.,* 1977a). Functional attributes of communities in an ecosystem, as opposed to structural attributes, are most frequently utilized for this purpose. Measurements of functional responses, although reflective of general ecosystem behavior, are quite often mechanistically based on the activities of microbial communities. They are typically community responses that theoretically can be attributed to a variety of ecosystem community structures and trophic interactions. In other words, they are measures of system behavior that empirical studies have shown to be indicative of ecosystem well-being and comparability.

Primary productivity and respiration are key functional parameters of most aquatic and terrestrial ecosystems. Substantial evidence now exists for aquatic systems that support the concept that the ratio of production (P) to respiration (R) is an excellent measure of system behavior and is consistent, sensitive to man-made perturbations, and easily compared in both laboratory and field situations (Giddings and Eddleman, 1978; Beyers, 1964; Cooke, 1974). Odum (1969) suggested that a P/R ratio of 1 can be attributed to mature ecosystems; this has been observed in both microcosms and natural systems. A reduction in the ratio signifies a negative energy balance; the particular system does not maintain itself and will continue to show a decrease in P/R ratio if no allochthonous or autochthonous energy sources are provided. A measurement of P/R ratio in an aquatic microcosm will consequently provide some indication of its self-sustaining characteristics relative to the field. It is thus reasonable to assume that a microcosm that tracks the P/R ratio of the field is likely to track other natural microbial processes in the field as well.

Nutrient cycling is, from a microbial community standpoint, probably the best field calibration checkpoint available. Dissolved inorganic nutrient concentrations in the water column or interstitial water of many aquatic systems are typically quite low due to the close coupling of remineralization rates, plant uptake rates, and physical exchange rates between water and sediment. A microcosm that shows higher water concentrations relative to the field indicates that this couple may not be functioning realistically and that the laboratory system is possibly not self-sustaining. The extent to which the response of microbial populations to a pollutant will be affected is uncertain, but it is generally believed that a good microcosm simulation should at least mimic nutrient cycling events in the field. Examples of how field calibration with nutrient cycling can be assessed are found in the nitrogen and sulfur cycling studies of Henriksen (1980) and Cappenberg (1974), respectively.

Biogeochemical cycles in soil and sediment are also closely linked to proton and electron activity or redox potential (Eh). Eh is an integrated measure of many chemical processes and redox reactions. Disturbing the redox gradient, as discussed previously, can lead to significant changes in metabolic rates of carbon turnover and mineralization by certain microbial communities. Consequently, a microcosm that does not model natural redox conditions is likely to provide a poor field simulation. Redox potential can be measured either by metallic probes (Waide *et al.*, 1980), by oxygen uptake (Edberg and Hofsten, 1973; Pamatmat, 1971), or by assays for dehydrogenase or electron-transport system (ETS) activity (Olanczuk-Neyman and Vosjan, 1977). In each case, good correlation between the microcosm (usually an intact core) and the field has been observed. Figure 18 shows how a tidal flat microcosm was field calibrated, using vertical distribution measurements of ETS activity, oxygen uptake, and sulfide concentrations (Vosjan and Olanczuk-Neyman, 1977). Similar measurements could also be made in terrestrial systems.

The measurement of dissolved organic carbon (DOC) concentrations resulting from secretions by living organisms and partial decomposition of dead organisms is another potential field calibration measurement. Transformation of organic carbon and the subsequent release of nutrients are major growth-controlling factors for the autotrophic community. As such, DOC would be expected to vary as a function of both autotrophic and heterotrophic activity. Burney *et al.* (1981) compared the flux of DOC and polysaccharide carbon (CHO) in a saltmarsh and a large outdoor tank and found not only significant differences between the tank and the field but also a strong relationship between bacterial activity and the DOC or CHO levels. They suggested that bacteria may respond rapidly to and control the concentrations of CHO and perhaps other types of DOC. Similar conclusions were reached by Larsson and Hadstrom (1979), Wiebe and Smith (1977), and Straskrabova and Fuksa (1982). Simulation of DOC turnover in a microcosm may therefore be a very sensitive field calibration method.

Concentrations of DOC also reflect the eutrophic condition of a natural system (Wangersky, 1978). Since a microcosm will contain excessive wall surface area for algal growth, DOC may be considerably elevated over natural conditions. This could have a significant effect on metabolic turnover by microbial communities and consequently on the fate of a pollutant. A misleading evaluation of risk assessment might therefore result if DOC and the associated bacterial activities are not calibrated with the field.

The measurement of CO_2 evolution rates from soil is a tested and reliable technique for assessing general biological activity. It is also a sensitive measurement that is often affected by sampling procedures and the subsequent incubation conditions. As with the other field calibration techniques, CO_2 evolution is an integrative measure that does not distinguish the relative contribution of the different living components in the sample, i.e., microorganisms,

I DUTCH WADDEN SEA

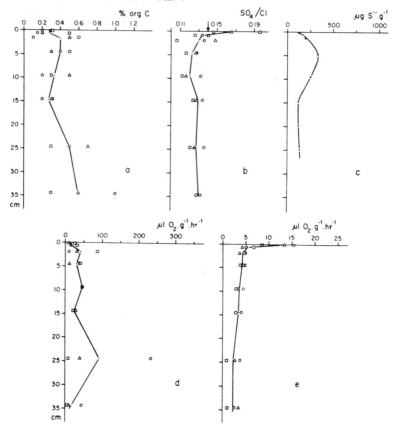

Figure 18. Vertical distributions in a natural tidal flat sediment from the Dutch Wadden Sea (I) and an artificial tidal flat system (II). (a) Organic carbon content (% of dry sediment); (b) sulfate/chlorinity ratio in the interstitial water [(↓) seawater ratio]; (c) sulfide content

plant roots, and invertebrate animals. Ausmus *et al.,* (1978) have shown the usefulness of this technique in comparing the litter decomposition behavior of a "treecosm" with the field under similar regimes of heavy-metal exposure.

9.2. Structural Attributes

Structural features of a microbial community may also provide information useful in calibrating a microcosm with the field. Although it is a delicate and time-consuming undertaking to describe a bacterial community according to certain species-diversity indices, it is feasible under certain circumstances.

II ARTIFICIAL TIDAL FLAT

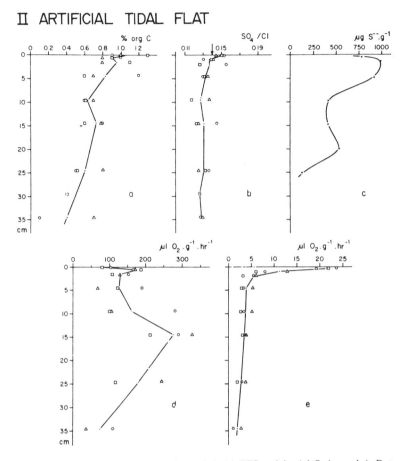

(μg S^2/g); (d) oxygen utilization (μl O$_2$/g per hr); (e) ETS activity (μl O$_2$/g per hr). Determinations in three separate cores; lines connect mean values. After Vosjan and Olanczuk-Neyman (1977).

Taxonomy of bacteria, because of poor morphological differentiation, commonly relies on the physiological, metabolic, or biochemical characteristics of bacteria to determine species composition and diversity. Troussellier and Legendre (1981) have suggested a unique "functional evenness index" that is based directly on the results of the API 208 testing method. The availability of this testing method would make the application of an evenness index feasible in calibrating a microcosm. Hauxhurst et al. (1981) have developed indices of physiological tolerance and nutritional utilization that could be used in a microcosm study to be certain that the natural diversity of the microbial communities was maintained. The nutritional utilization indices in this study were based on the qualitative parameter of presence or absence of growth on a large variety of single carbon source media. Similar indices could probably be established for the quantitative utilization of a substrate, using standard ^{14}C heterotrophic activity measurements. Goulder et al. (1978, 1980), for example, have shown that the V_{max}/bacterium for glucose mineralization in estuarine water samples was negatively correlated with industrial pollution, particularly stresses due to sewage and copper. However, if the containment of a microbial population in a microcosm is considered stressful in itself, the resulting heterotrophic indices could be a good indication of field simulation.

White et al. (1979) have pioneered a biochemical approach of assessing prokaryotic and eukaryotic biomass and community structure in natural systems in which a qualitative analysis of lipid and fatty acid compositions in a sample is used. This approach is an excellent means of comparing sediment microbial community structure (Bobbie and White, 1980) in natural ecosystems and microcosms. For example, most bacteria contain diacylphospholipids. Actinomycetes, corynebacteria, and most eukaryotic protists contain a high proportion of ether-, vinyl ether-, and amide-phospholipids. Anaerobic bacteria exclusively contain plasmalogens. Short-chain (13–21 carbons) iso- and anteisobranched fatty acids are typical of gram-positive bacteria, and cyclopropane fatty acids are largely associated with gram-negative bacteria (White et al., 1979). If the relative composition of these compounds in an environmental sample is unique, then tracking of the microcosm biochemically with the field is possible. Various kinds of drilling-mud components have been shown to alter the lipid and fatty acid composition of sand colonized by the biota from flowing seawater (Smith et al., 1982).

There are many other potential field calibration measures available. Almost any general parameter of bacterial community structure and function that is sensitive to various kinds of stresses is applicable to microcosm design and operation, because it can be assumed that a microcosm behavior that deviates from that observed in the field is for all practical purposes stressed. Adenylate energy charge (Chapman and Atkinson, 1977), a variety of enzymatic activities (Griffiths et al., 1982), especially chitinase activity (Portier and Meyers, 1981), the microbial composition of the sea surface microlayer (Hardy and Valett, 1981), and of course a whole range of heterotrophic activity measurements are all possible field calibration measures.

9.3. Fate Studies

A final type of field calibration approach is to examine the handful of cases in which the fate of a particular chemical in a microcosm has been compared with a controlled field site. Several studies indicate differences in biodegradation rates between the laboratory and the field; such examples could be examined further to determine what factors in microcosm design or operation were responsible for these differences. Sikka and Rice (1973), for example, compared the fate of the herbicide endothall in a farm pond and a laboratory aquarium microcosm. Their results are shown in Fig. 19. Overall, the pattern of distribution (partitioning to hydrosoil) and disappearance of parent compound was the same, but the magnitude of these processes was considerably different. Endothall accumulation in the microcosm sediments was approximately one half that found in pond sediments; the highest concentration of herbicide in sediment occurred much earlier (a difference of 20 days) in the microcosm, and the loss rate of parent compound from the sediment after peak concentration was much more rapid in the pond. Degradation in the water column was much faster in the microcosm due primarily to a shorter lag in an apparent adaptation to greater biodegradation rates. The reduced lag may have resulted because the microcosms were established originally with pond water and sediment taken from a control pond that was different from the dosed pond. The latter had been previously (12 months earlier) exposed to endothall (Sikka and Rice, 1973). The other differences might be attributable to an approximate 9-fold difference in sediment surface area/water volume ratio between pond and microcosm and possibly slow mixing in the pond. The potential for calibrating the microcosm with the pond is apparent.

Tsushimoto *et al.* (1982) compared the fate of 2,3,7,8-tetrachlorodibenzo-*p*-dioxin (TCDD) in an outdoor pond (man-made) and in a 16-cm-diameter glass bottle microcosms. Data from each test system were roughly comparable for short-term distribution behavior. However, a large die-off of the pond weeds *Elodea* and *Ceratophyllon* in the outdoor pond caused a significant change in the overall physical distribution of the TCDD, probably because of large increases in sorptive surface competing with similar surfaces on the sediment. It is not clear whether a similar shift would have occurred in the microcosm study, since a pond weed die-off did not occur with the shorter incubation period. Significant distributional differences may have resulted if plant die-off had occurred in the microcosm. The relative proportions of sediment, water, and plant components in the microcosm were not calibrated with the pond.

A related microcosm study involving the effect of plant material on the fate of diquat also indicates the need for field calibration. Simsiman and Chesters (1976), using flasks containing static sediment–water components as simulated lake impoundments, showed a significant depression in diquat degradation because of oxygen depletion in the system caused by the death and decay of plant material. Their suggestion that this may be a potential result in the field is very tenuous, considering the disproportionate ratios of water volume to plant mass between the flask and any field situation.

Kloskowski *et al.* (1981) have reported a study comparing the degradation (including conversion to bound residues and carbon dioxide) of 11 different chemicals in a soil–plant (summer barley) microcosm to that occurring in an outdoor experimental area. The relative persistence of the chemicals was similar in both laboratory and field tests, but the magnitude of degradation in each

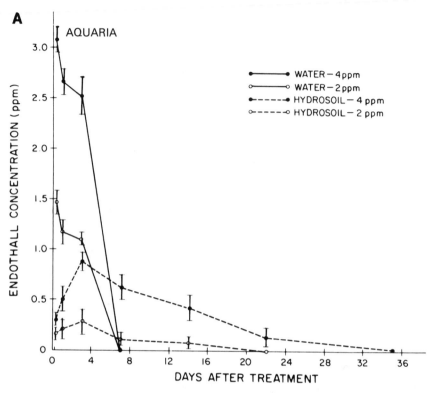

Figure 19. Endothall residues in water and hydrosoil taken from laboratory aquaria (A) and from a farm pond (B). The aquaria were treated with 2 and 4 ppm of the herbicide and the farm pond with 2 ppm. (I) Ranges of duplicate values. After Sikka and Rice (1973).

test situation differed considerably for certain chemicals. Unfortunately, soil type and plant species in the field plots were not the same as those in the microcosm; consequently, it is not certain how these factors affected the extents of degradation. Elner *et al.* (1981) have studied the movement of a carbamate herbicide, aminocarb, into the surface microlayer of the water column of a tank system and an experimental outdoor pond and found comparable rates of biodegradation following aerial application of the pesticide. Carbofuran, a systemic insecticide, was found to be degraded slowly in laboratory soil systems (Ahmad *et al.,* 1979), leading the authors to conclude that rapid degradation was not the reason for the ineffective control of corn rootworm infestation in one particular soil. Without field calibration of the laboratory system, however, questions arise about the simulation of actual field conditions in the laboratory system, and their conclusion may not be correct.

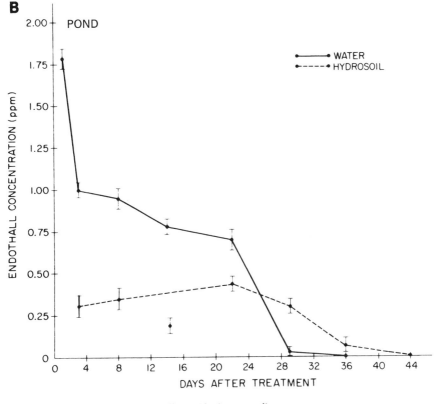

Figure 19. (*continued*)

10. Precautions in the Use of Microcosms

Although many of the virtues of using microcosms in microbiological research have been discussed, potential limitations to microcosm use should be considered, particularly in relation to studying the microbial responses to pollutants. Probably the most obvious problem in microcosm technology is scaling; i.e., how does the isolation of a piece of an ecosystem affect the function of microbial communities, or as smaller and smaller pieces of an ecosystem (scale reduction) are isolated, how many sites within that ecosystem must be sampled before their summarized response or activity equals the whole? There are basically four aspects to this scaling problem: process uncoupling, linear process control, artificial surface enhancement, and specialized compartments within ecosystems. Each poses a problem that may not necessarily be correctable in the laboratory, but their potential effects on the interpretation of microcosm results should be carefully considered. Certainly, microbiologists are far less likely to experience effects due to scaling because microbial populations are so metabolically diverse and, due to rapid growth, are able to compensate for losses or changes in community structure.

10.1. Process Uncoupling

The fundamental thermodynamic basis of ecosystems involves the generation of energy from the microbial oxidation of degradable organic materials and the consequent mobilization of inorganic nutrients produced from primary productivity and geochemical events external to the ecosystem. Ecosystems are viewed as biogeochemical units reflecting the integration of a large variety of metabolic networks. A microcosm will therefore incorporate a certain proportion of these networks, but not all (Schindler *et al.*, 1980). This is diagrammatically shown in Fig. 20. A particular microcosm will include a certain amount of the biogeochemistry of an ecosystem (sizes of the microcosm circles), and it will reflect a certain amount of the true dynamics and interactions of an ecosystem (overlap into the ecosystem circle). For most microcosm studies, the circle will be large, indicating an accommodation of many of the metabolic networks found in nature. However, there will be lesser degrees of overlap, indicating limitations or exaggeration of the dynamics of these networks in the microcosm.

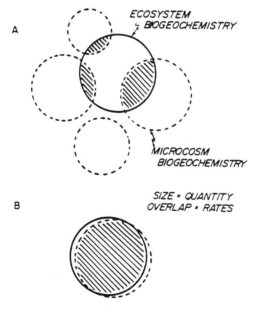

Figure 20. Schematic diagram portraying possible interrelationship between the biogeochemistry of an ecosystem and various microcosms. The sizes of the microcosm circles (– – – –) indicate the extent to which they incorporate the biogeochemistry of the ecosystem (————). Overlap of the circles indicates similarities in rates of biogeochemical processes in microcosms and ecosystem. (A, B) Possible examples of these interrelationships.

Rarely would a situation exist as shown in Fig. 20B, in which a single microcosm would account, qualitatively and quantitatively, for all the biogeochemistry of the ecosystem. Arbitrary boundary conditions established for an environmental sample, as often occurs in a microcosm, uncouple some of the biogeochemical processes from the ecosystem. Uncoupling may not eliminate any critical components, or change the number and types of microbial interactions, but it could greatly upset the *rates* at which these interactions occur. A common example of process uncoupling in microcosms occurs when higher trophic levels, such as plankton and grazer communities, are excluded. Goodyear *et al.* (1972) and Jassby *et al.* (1977b) have shown, for example, that grazing by herbivorous fish greatly stimulates primary productivity. Flint and Goldman (1975) showed similar responses of diatom communities to grazing by crayfish in freshwater lakes. Laboratory systems excluding the grazers would give conservative estimates of plankton activity.

A number of studies have shown that bacteriovorous animals directly and indirectly affect the growth and metabolic activity (O_2 uptake, nutrient cycling, and decomposition rates) of bacterial populations (Fenchel and Harrison, 1976; Fry, 1982). Hargrave (1970) showed that bacterial respiration in lake sediment cores increased with increasing numbers of the freshwater amphipod *Hyaletta azteca*. Barsdate *et al.* (1974) demonstrated that certain protozoa, while causing reductions in bacterial density, stimulated the metabolic activity of the bacteria as measured by increased rates of phosphorus uptake. The deposition and processing of animal fecal material are also crucial to the activity of microorganisms in sediment, presumably providing a source of nitrogen (Hargrave, 1976; Lopez *et al.*, 1977). Several investigators have shown that the grazing of detritus-associated bacteria by protozoa and amphipods stimulates the bacterial mineralization of organic detrital material (Barsdate *et al.*, 1974; Harrison and Mann, 1975; Lopez *et al.*, 1977). A typical result using radiolabeled hay as detritus in a seawater system is shown in Fig. 21. Similar types of relationships between grazing organisms and the activity of microbial populations have also been demonstrated in terrestrial systems (Elliott *et al.*, 1979;

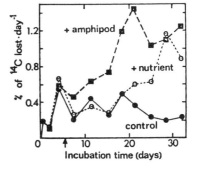

Figure 21. Daily rate of loss of ^{14}C (as $^{14}CO_2$) from labeled hay incubated at 5°C in seawater, showing the effects of addition of inorganic nutrients or of an amphipod at the time shown by the arrow under the abscissa. Points are means of two replicates. After Harrison (1977).

Stout, 1980). The exclusion of any of these trophic components in a microcosm could therefore significantly alter natural metabolic rates of microbial communities.

In addition, burrowing benthic invertebrates, in the process of reworking sediments and increasing the size of the oxidized zone, can cause increased microbial activity (Hylleberg and Henriksen, 1980). Martin and Bianchi (1980), Burney et al. (1981), Straskrabova and Fuksa (1982), and McKinley and Wetzel (1979) have also shown that the metabolic activity of aquatic bacteria is closely coupled to the release of organic materials by growing populations of phytoplankton.

These examples point out that microbially mediated processes in sediment are integrally coupled with a variety of trophic levels in an ecosystem. As discussed already, these couplings may be very sensitive points in the degradative and toxicological response of microbial populations to pollutants.

A good microcosm study should accommodate as many trophic level interactions as possible. In some cases, the size or restrictions of microcosms in the laboratory will preclude the presence of certain trophic levels, particularly fish species and large invertebrates that are not only difficult to maintain in microcosms (due to poor renewal of natural food sources) but also voracious feeders, often totally denuding the physical structure of the microcosm components. Gillett and Gile (1976), for example, found that a meadow vole in a terrestrial synthetic community study totally consumed all the vegetation and completely turned over all the soil. Brockway et al. (1979), using pond microcosms, found that snails introduced functional instability to the system and had to be removed. The only way to accommodate some of these higher trophic levels, other than using impractical large tanks, is to add these macrofauna to self-sustaining microcosms for only short experimental periods and then observe the changes in function and structure of the microcosm ecosystem as though they were a zero-order effect. In other cases, intact cores (from 3 to 30 cm in diameter) from soil or sediment will naturally contain large numbers of macrofaunal organisms. A mixed-culture study using sediment slurries in a shake flask cannot accommodate the important biogeochemical couples with different trophic levels.

A second type of process uncoupling involves physical rather than biological factors. Large bodies of water, for example, experience physical mixing and turbulence phenomena that are difficult to model in laboratory tanks and are generally ignored. They may have a significant effect, however, on the process rates of microbial communities. Thermoclines in freshwater lakes are difficult to simulate in microcosms (Harte et al., 1980). It is therefore very complicated in the laboratory to investigate the interchange of inorganic nutrients, microorganisms, and organic substrates among the hypolimnion and the epilimnion, as well as the profundal and littoral zones. Isolating the epilimnion in a microcosm (Jassby et al., 1977a), without taking into consideration the inputs from the hypolimnion, could produce misleading rate information about microbial processes contained therein. Dissolved oxygen concentration in the hypolimnion, for example, is inversely related to the distance from the sediment (J. G. Jones, 1982). Modeling each zone separately in a microcosm and studying output responses from a variety of inputs, including pollutants, may provide enough information to model the interactions and feedbacks of the zones mathematically. J. G. Jones (1982) has reviewed some of the sediment-coring methods that could be used in these microcosm studies.

Actual turbulence and water velocity of lotic environments are often difficult to accommodate in microcosms, and their absence may affect the natural behavior of microbial communities. Leaf litter is a dominant component of the coarse particulate organic matter in woodland streams. It commonly accumulates at the leading edge of various obstructions, i.e., rocks, wood, and debris jams, where it forms into packs due to the current. The processing of these leaf packs into fine particulate organic matter occurs through physical leaching and the combined action of "shredders" in the invertebrate communities and decomposition by colonizing microorganisms (Peterson and Cummins, 1974). Stream current is important in controlling the leaf processing rate, since it supplies oxygen, continuously adds new matter, removes decomposed materials, and provides a proper habitat for the invertebrates. The absence of a current in a laboratory microcosm leads to the development of invertebrate and bacterial communities more typical of pool areas in a stream where litter decomposition is much slower (Cummins, 1974). In a similar situation, McIntire *et al.* (1975) have summarized evidence that demonstrates the dependence of primary productivity in periphyton assemblages associated with rock surfaces on the stream current. For example, periphyton productivity is very susceptible to the exhaustion of certain dissolved gases and nutrients; the gradients established for these materials are generally controlled by stream turbulence (McIntire, 1966). Likewise, dislodgment of periphyton from rocks, possibly by the scouring action of suspended solids, greatly promotes respiration and may account for the resulting higher periphyton biomasses in rapidly flowing streams (McIntire, 1978). It is apparent, therefore, that any microcosm design that ignores these mechanical features will likely give a false picture of natural process rates.

Some of the best and most extensively tested microcosms have been developed to study stream ecology. In many cases, adequate simulation of currents and turbulence has been mechanically achieved, providing ecosystem behavior in excellent agreement with field measurements (McIntire, 1964; Gee and Bartnik, 1969; Lauff and Cummins, 1964; Warren and Davis, 1971; Pritchard and Cripe, 1983). These studies have provided ideal, but largely untested, laboratory systems for assessing the relationship between the fate and effects of pollutants and the activity of microbial communities. S. R. Hansen and Garton (1982) and Maki (1980) have shown their potential use in assessing the toxicological effects of the pesticides diflubenzuron (a chitin-synthesis inhibitor) and trifluoromethyl-*p*-nitrophenol (a lampricide), respectively. Giesy (1978) has demonstrated, using a microcosm, the toxicity of cadmium to leaf litter colonization of bacteria and fungi and to its eventual decomposition. Bott *et al.* (1976) have examined the fate of nitrilotriacetic acid in microcosms. In each example, modeling of stream current and turbulence was critical to the environmental significance of the results.

In their excellent series of studies with marine microcosms and the dynamics of plankton communities, Perez *et al.* (1977) and Nixon *et al.* (1979) have shown that light intensity and turbulence are important in determining plankton biomass and in realistically modeling the couple between benthic and pelagic phases. Turbulent mixing of the water in their 150-liter tanks was obtained with specially designed paddles made of a plastic grid (12 mm $\times$ 12 mm openings) and driven by electric motors that reversed direction every 30 seconds. Water was also passed across the sediment surface (contained in a plastic box) at a rate that exposed all the water in the tank to the sediment once every 6 days and thereby eliminated any functional gradients. Despite considerable precautions, the microcosm generally underestimated turbulent mixing in the field. It is therefore not a trivial matter in a pelagic model of this type to simulate natural mixing and turbulence in a microcosm. Depending on the research question being addressed, ignorance of this design feature may result in process rates that are physically uncoupled from normal ecosystem behavior. It may also significantly affect the distribution of toxicant and, correspondingly, the ultimate fate and effects of a pollutant.

Natural lighting is also a complicated physical factor to model in the laboratory systems. As indicated before, the phytoplankton, epiphyte, and rooted macrophyte communities in many types of ecosystems can influence the metabolic rates of microbial communities. This is due to their excretion of organic material, their contribution to decomposable organic detritus, and their role as a food source for zooplankton and benthic macrofauna that elaborate and process degradable fecal material. These plant communities should therefore be included in microcosms. Standard, cool-white 40 W fluorescent lights are commonly used to maintain the plant communities. However, Nixon *et al.* (1979) have shown that light intensities in the microcosm that approximate those seen in the field (Narragansett Bay, Rhode Island) cause "super blooms" of phytoplankton (Fig. 22). Intensities of 5 langleys (1y) per day in the microcosm produced algal populations similar to those observed in the field. However, light intensities of around 40 ly/day were observed in the field, and intensities greater than 40 ly/day initiated an algal bloom therein. Interestingly, bloom conditions in the microcosms were considerably more dramatic than in the field; "instantaneous" growth rates of 0.23/day (25 ly/day) and 0.03/day (>40 ly/day) were obtained, respectively. Explanations for these differences may lie in the spectral quality of natural light and its absorption characteristics

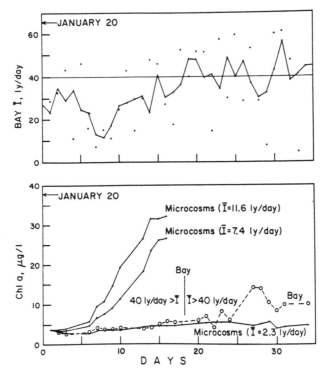

Figure 22. Chlorophyll *a* concentrations over time in the lower west passage of Narragansett Bay and in laboratory microcosms maintained at different light levels but with the same photoperiod as the bay. The microcosm temperatures were also kept within 1–2°C of the bay temperature. Results for the microcosms at light intensities of 11.6, 7.4, and 2.3 ly/day are shown. Values plotted are the means for duplicate tanks. Daily values (●) for the approximate mean light in the water column in the bay are shown, along with the 3-day moving average (---) for the period preceding and during the bay bloom in February 1975. After Nixon *et al.* (1979).

in natural waters. Scaling of light intensity and quality (spectral distribution) is therefore essential if the natural activities of plant communities and their dynamic interactions with microbial communities are to be studied and perhaps related to pollutant perturbations.

A final microcosm design feature that can cause process uncoupling is the use of a closed microcosm when the environmental situation being modeled is quite open. In closed systems, microcosms are ecosystems unto themselves, having self-sustaining properties and nutrient-cycling dynamics that are unique environmentally but conceptually within the realm of the general ecosystem theory (Hill and Wiegert, 1980). This concept may be true except that in a closed system, the equilibrium state will be a function of the initial conditions. The chance that this equilibrium state will be equivalent to some environmental counterpart is remote. Process rates may therefore be greatly exaggerated (faster or slower) and essentially uncoupled from the natural ecosystem function. In an open system, however, the equilibrium state is a function of the inputs and outputs, and regardless of which portion of an ecosystem is modeled in a microcosm, it must experience a set of inputs and outputs similar to what it would receive in the field. This consideration of inputs and outputs in microcosm study is unfortunately ignored in many cases. Inputs, typical field conditions, can often be modeled by continuous-flow systems in which the water source is taken directly from the environment under study. Thus, a continual input of microbial, algal, and invertebrate larval biomass can be achieved in the laboratory. Organic and inorganic nutrient inputs and outputs can also be modeled, thereby preventing growth limitation due to nutrient depletion or to the buildup of toxic end products that would normally be associated with a closed system. Lassiter (1979) has proposed that continuous-culture theory be applied to open-microcosm studies. DePinto et al. (1980) have demonstrated the value of a continuous-flow system relative to a closed system in studying techniques for acid lake recovery.

10.2. Linear Process Control

As stated earlier, one of the criteria for applying process rate data from mixed-culture studies of microbial communities in the laboratory to actual field situations is the verification exercise showing the applicability of the microcosm data to the field. It is then logical to ask how generally applicable this verification is. Can similar results be obtained with other compounds *and* in other types and designs of microcosms when considering other parts of the same ecosystem, or similar ecosystems of different geographic location, or completely different ecosystems? The development of kinetic expressions for biodegradation provides a relevant example. Assume that a chemical degrades much more rapidly in a shake-flask test containing a sediment–water slurry than it does in one containing only water. Since it is often difficult to characterize quantitatively the degrading microbial populations associated with sediment, the development of kinetic expressions involving rates per cell is likewise difficult. As a surrogate, the degradation rate can be related to the amount of organic detritus available, assuming that the distribution of bacteria on each sediment particle is approximately the same. This is possible for the degradation of a number of pesticides in sediment and soil where a biologically mediated binding to particulates is the principal degradation process (Katan et al., 1976; Pritchard et al., 1979; Sethunathan et al., 1977). A kinetic expression relating binding rate to the mass of sediment, which has been successfully used to describe the biological fate of a pesticide in a sediment–water microcosm (Pritchard et al., 1983b), means that with a sediment sample from any similar environment, the biodegradation rate can be predicted by knowing the average sediment concentration and the binding rate constant. This assumes that the relationship of sediment concentration to degradation rate is linear over a large range of detritus concentrations. Studies in our laboratory (Cleveland, 1982) have shown that the realtionship for estuarine sediments is not entirely linear (Fig. 23). Binding of methyl parathion to three different sediment types was repressed to different degrees at high sediment concentrations, indicating that active sites were covered, perhaps due to clumping. Verification of the binding rate expression in microcosms was obtained with the Range Point sediment (Pritchard *et al.*, 1983b). The question of which type of environmental situation or, since the environment cannot be dosed with the chemical, which type or size of microcosm is required before the effect of this nonlinear relationship can be perceived and verified must be answered. The answer is not obvious and may come only from a large set of empirical experiments, such as testing a whole gradation in size and complexity of microcosms.

In most research settings, scaling experiments are not possible. It is enough to be able to adequately describe the events in a single microcosm. However, some scaling work will eventually be necessary, and it is hoped that

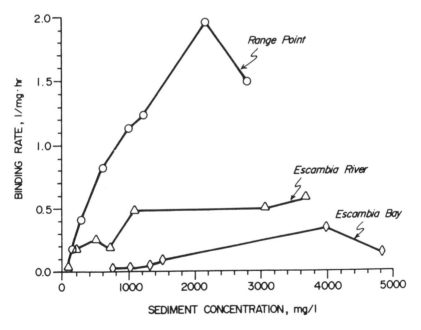

Figure 23. Effect of sediment concentration on binding of ^{14}C (ring-labeled) methyl parathion to three different sediment types taken from the Florida Gulf Coast. Binding is the amount of radioactivity that becomes associated with sediment and cannot be removed by solvent extraction. The bound radioactivity is measured by combustion and trapping of $^{14}CO_2$. Binding experiments were carried out in shake flasks in which sediment slurries were completely mixed. After Cleveland (1983).

results from a few experiments will have general application. In the meantime, careful attention must be paid to checking the linearity of kinetic expressions to at least become aware of where problems in extrapolation might be expected.

10.3. Wall Effects

The growth of bacteria and other epiphytes on the walls of the microcosm container can potentially have dramatic effects on the experimental results. In many microcosm studies, this problem is unfortunately overlooked or in many cases simply ignored. Generally, within hours to days, a microbial film begins to develop on the walls of any container used to house natural water or sediment samples. A succession of bacterial types usually results, and colonization by protozoa, diatoms, and other algae generally follows. Long-term incubations, typical of many microcosm studies, can produce enough wall growth to represent a substantial part of the biology in the microcosm. Dudzik et al. (1979), for example, have calculated that a close-packed 1-μm layer of algal cells on the walls of a microcosm tank 1 m in diameter and 1 m deep has about the same volume density as phytoplankton in a small mesotrophic lake! Release of metabolic end products, including polymeric materials, death and decay of cells, and sorption of organic and inorganic materials are all consequences of wall growth that can greatly influence carbon and nutrient cycling rates in a microcosm.

Several approaches can be used to accommodate the wall-growth phenomenon. It definitely cannot be ignored. Continual scraping of the vessel walls is a feasible approach (Perez et al., 1977), but this procedure often leads to unnaturally high concentrations of suspended solids or the deposition of large amounts of degradable organic material on the sediment surface. Depending on the research question being asked, the effects of this additional organic loading must be reconciled in the interpretation of microcosm results. Harte et al. (1980) have surmounted the wall-growth problem by periodically transferring the contents of a microcosm to a clean vessel. Although this is potentially feasible in a microcosm designed to model the epilimnion of a lake, it is not very practical for systems with sediment or with other highly structured components.

A more reasonable approach, in the authors' opinion, is to accept the wall growth as inevitable and then determine its role in the biology and nutrient cycling in the microcosm. This could be accomplished by first adjusting the operating conditions of the microcosm in an attempt to possibly slow the rate of wall-growth accumulation. Increasing the flow rate of water through the system (if this is possible) or increasing turbulent mixing might attain this result. Second, by hanging glass or Plexiglas slides in the microcosm, it should be possible to surrogately sample the wall growth and examine or measure the associated microbial activity. These results can then be quantitatively factored into the interpretation of the microcosm results.

10.4. Specialized Compartments within Ecosystems

The object of a microcosm study, as already discussed, is to capture certain portions of the ecosystem that the investigator feels could be critically important in the overall assessment of the response of a microbial community to a pollutant. What constitutes or defines a "portion" of the ecosystem? This is a complicated question that will affect assumptions in designing a microcosm and will cause problems in extrapolation. Take, for example, the cross section of an estuarine ecosystem shown in Fig. 24. An assumption in modeling the ecosystem in a microcosm may be to establish a sediment–water system representing the average sediment surface area/water volume ratio of the bay. This assumption may be tenable for some studies, but from the standpoint of exposure assessment in which a chemical may biodegrade much more rapidly in sediment than in water, it is not tenable, since the chemical will disappear at a faster rate in shallower waters where it encounters sediment more frequently. Thus, the microcosm results may give a conservative estimate of fate because of this original assumption.

To elaborate further, biodegradation may also be more rapid in and around rooted macrophytes, such as are found in grass flat areas and adjoining saltmarshes. Higher concentrations and activities of bacteria on the plant leaf surfaces, in the root rhizosphere, and in the associated detrital material may contribute significantly to the biodegradation of certain chemicals. This component of pollutant biodegradation has been virtually ignored over the years, yet its potential contribution would seem to merit further experimental consideration, particularly in microcosm studies. Hsu and Bartha (1979) have shown in soil systems that bacterial populations associated with the roots of plants degrade parathion faster than populations from areas without these plants. A microcosm designed to model only the more geographically dominant neritic component of the estuarine system would therefore ignore this contribution from the plants. Thus, if this type of microcosm is used to study the fate of a chemical in an estuary, estimation of the fate of the chemical in the whole ecosystem would be erroneous. Other areas of estuaries would also appear to

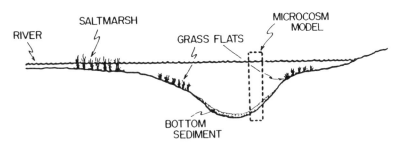

Figure 24. Diagrammatic cross section of a typical estuarine ecosystem. The dashed-outline box represents the portion of the ecosystem that could be modeled in a microcosm.

be unique: transition zones between fresh and salt water, where sediment resuspension could greatly affect biodegradation rates, and certain embayments where different organic and nutrient loadings, including anthropogenic sources, could affect fate processes significantly. Similar considerations can be cited for lakes, rivers, and terrestrial environments.

Where does one draw the line? How many microcosm studies will be needed to adequately describe the response of microbial communities to a pollutant, especially in aquatic systems where exposure is maximized by water mixing? There are three potential solutions to this problem. First, a hydrodynamic mathematical model of the particular environment of concern can be used to estimate the residence time of a chemical in a perceptually unique part of an environment. The EXAMS model (Burns et al., 1982) and the WASTOX model (Connolly and O'Connor, 1982) are excellent generalized models that can be applied to a large variety of aquatic systems. If the estimated residence time in one of the subsidiary parts of an ecosystem (a grass bed area, for example) is short relative to a major part of the ecosystem (the water column), then even though the former part may have a considerably greater biodegradation potential than the latter, its effect will go unnoticed because of hydrodynamic conditions of the environment. Consequently, the number of unique parts of an ecosystem that need to be considered for experimentation can be reduced.

A second solution would require actual experiments with each perceived unique part of an ecosystem in microcosm studies, establishing a ranking of parts that had the most significant responses of the microbial populations, and then assuming that their relationship to one another remained constant from one ecosystem to the next. For example, bacterial populations associated with plant roots might stimulate the degradation of a toxicant by a factor of only 2 relative to the activities in sediment. This is not significant enough in an estuarine ecosystem to merit further consideration, since the relative mass of sediment and its firsthand exposure to the pollutant are considerably higher. On the other hand, a factor of 5–10 would probably be significant. Once this relationship is established, there will probably be few exceptions. The overall contribution of each significantly active part to the ecosystem can again be assessed through hydrodynamic modeling.

Finally, a third solution, which is perhaps the most difficult but the most useful, is to characterize the factors that make one unique part of an ecosystem more microbially active than another. This type of characterization would come from a combination of microcosm and mixed-culture tests as already discussed. For example, if an observed increase in biodegradability of a pollutant associated with the microbiota of rooted macrophytes could be attributed to larger degrader populations, then if the biodegradation rate constant (degradation rate per cell) of a pollutant and the concentration of degraders in each particular unique part of an ecosystem are known, the biotic fate of a pollutant can be assessed. This does not really reduce the number of initial microcosm studies, but it allows the consistency of the relationship to be checked from ecosystem to ecosystem. More important, it eliminates the need to conduct microcosm studies in several sites within a unique part of the ecosystem as a check on the variability of the recorded responses.

References

Abbott, W., 1966, Microcosm studies on estuarine waters. I. The replicability ofmicrocosms, *J. Water Pollut. Control Fed.* **38**:258–270.

Adler, D., Amdurer, M., and Santschi, P. H., 1980, Metal tracers in two marine microcosms: Sensitivity to scale and configuration, in: *Microcosms in Ecological Research* (J. P. Giesey, ed.), pp. 348–368, Department of Energy, Symposium Ser. 52 (Conf-781101), NTIS, Springfield, Virginia.

Ahmad, N., Walgenbach, D. D., and Sutler, G. R., 1979, Degradation rates of technical car-bofuran and a granular formulation in four soils with known insecticide use history, *Bull. Environ. Contam. Toxicol.* **23**:572–574.

Albright, L. J., and Wilson, E. M., 1974, Sublethal effects of several metallic salts–organic compound combinations upon the heterotrophic microflora of a natural water, *Water Res.* **8**:101–105.

Anderson, J. R., 1978, Pesticide effects on non-target soil microorganisms, in: *Pesticide Microbiology* (I. R. Hill and S. J. L. Wright, eds.), pp. 313–359, Academic Press, London.

Anderson, R. V., Elliott, E. T., McClellan, J. F., Coleman, D. C., Cole, C. V., and Hund, H. W., 1978, Trophic interactions in soils as they affect energy and nutrient dynamics. III. Biotic interactions of bacteria, amoebae, and nematodes, *Microb. Ecol.* **4**:361–371.

Atlas, R. M., Pramer, D., and Bartha, R., 1978, Assessment of pesticide effects on non-target soil microorganisms, *Soil Biol. Biochem.* **10**:231–239.

Ausmus, B. S., Dodson, G. J., and Jackson, D. R., 1978, Behavior of heavy metals in forest microcosms. III. Effects on litter-soil carbon metabolism, *Water Air Soil Pollut.* **10**: 19–26.

Babich, H., and G. Stotzky, 1980, Environmental factors that influence the toxicity of heavy metals and gaseous pollutants to microorganisms, *Crit. Rev. Microbiol.* **8**:99–145.

Barsdate, R. J., Prentki, T. R., and Fenchel, T., 1974, Phosphorus cycle of model ecosystems: Significance for decomposer food chains and effect of bacterial grazers, *Oikos* **25**:239–251.

Baughman, G. L., and Burns, L. A., 1980, Transport and transformation of chemicals: A perspective, in: *Handbook of Environmental Chemistry*, Vol. 2, Part A (O. Hutzinger, ed.), pp. 1–17, Springer-Verlag, Berlin.

Bazin, M. J., Saunders P. T., and Prosser, J. I., 1976, Models of microbial interactions in the soil, *Crit. Rev. Microbiol.* **4**:463–498.

Beers, J. R., Stewart, G. L., and Hoskins, K. D., 1977, Dynamics of micro-zooplankton populations treated with copper: Controlled ecosystem pollution experiment, *Bull. Mar. Sci.* **27**:66–79.

Beyers, R. J., 1964, The microcosm approach to ecosystem biology, *Am. Biol. Teach.* **26**:491–497.

Billen, G., 1982, Modeling the processes of organic matter degradation and nutrient recycling in sedimentary systems, in: *Sediment microbiology* (D. B. Nedwell and C. M. Brown, eds.), pp. 15–52, Academic Press, New York.

Billen, G., Joiris, C., and Wollast, R., 1974, A bacterial methyl mercury-mineralizing activity in river sediments, *Water Res.* **8**:219–255.

Bisogni, J. J., and Lawrence, A. W., 1975, Kinetics of mercury methylation in aerobic and anaerobic aquatic environments, *J. Water Pollut. Control Fed.* **47**:135–152.

Blau, G. E., and Neely, W. B., 1975, Mathematical model building with an application to determine the distribution of Dursban insecticide added to a simulated ecosystem, *Adv. Ecol. Res.* **9**:133–163.

Blum, J. E., and Bartha, R., 1980, Effect of salinity on methylation of mercury, *Bull. Environ. Contam. Toxicol.* **25**: 404–408.

Bobbie, R. J., and White, D. C., 1980, Characterization of benthic microbial community structure by high resolution gas chromatography of fatty acid methyl esters, *Appl. Environ. Microbiol.* **39**:1212–1222.

Bott, T. L., and Rogenmuser, K., 1978, Effects of No. 2 fuel oil, Nigerian crude oil, and used crankcase oil on attached algal communities: Acute and chronic toxicity of water-soluble constitutents, *Appl. Environ. Microbiol.* **36**:673–682.

Bott, T. L., Preslan, J., Finlay, J., and Brunker, R., 1976, The use of flowing-water microcosms and ecosystem streams to study microbial degradation of leaf litter and nitrilotriacetic acid, *Dev. Ind. Microbiol.* **18**:171–184.

Bourquin, A. W., and Pritchard, P. H. (eds.), 1979, *Workshop: Microbial Degradation of Pollutants in Marine Environments,* U.S. Environmental Protection Agency, EPA-600/9-72-012, Gulf Breeze, Florida, 552 pp.

Bourquin, A. W., Walker, W. W., and Pritchard, P. H., 1981, Screening test to estimate the degradation rates of toxicants in estuarine environments, *Am. Soc. Microbiol. (abstr.),* p. 276.

Boyle, T. P., Robinson-Wilson, E. F., Petty, J. D., and Weber, W., 1980, Degradation of pentachlorophenol in simulated lentic environment, *Bull. Environ. Contam. Toxicol.* **24**:177–184.

Branson, D. R., 1978, Predicting the fate of chemicals in the aquatic environment from laboratory data, in: *Estimating the Hazard of Chemical Substances to Aquatic Life* (J. Cairns, K. L. Dickson, and A. W. Maki, eds.), pp. 55–70, ASTM STP 657.

Brockway, D. L., Hill, J., Maudsley, J., and Lassiter, R. R., 1979, Development, replicability and modeling of naturally derived microcosms, *Int. J. Environ. Stud.* **13**:149–158.

Bull, A. T., 1980, Biodegradation: Some attitudes and strategies of microorganisms and microbiologists, in: *Contemporary Microbial Ecology* (D. C. Ellwood, J. N. Hedger, M. J. Latham, J. M. Lynch, and J. H. Slater, eds.), pp. 107–136, Academic Press, New York.

Burney, C. M., Johnson, K. M., and Sieburth, J. McN., 1981, Diel flux of dissolved carbohydrate in a salt marsh and a simulated estuarine ecosystem, *Mar. Biol.* **63**:175–187.

Burns, L. A., Cline, D. M., and Lassiter, R. L., 1982, *Exposure Analysis Modeling System (EXAMS): User Manual and System Documentation,* U.S. Environmental Protection Agency, EPA-600/3-82-023, 316 pp.

Cairns, J., 1981, Biological Monitoring. VI. Future needs, *Water Res.* **15**:941–952.

Cappenberg, T. E., 1974, Interrelations between sulfate-reducing and methane-producing bacteria in bottom deposits of a fresh-water lake. I. Field observations, *Antonie van Leeuwenhoek. J. Microbiol. Serol.* **40**:285–295.

Case, J. N., 1978, The engineering aspects of capturing a marine environment, CEPEX and others, *Rapp. P. V. Reun. Cons. Int. Explor. Mer.* **173**:49–58.

Chapman, A. W., and Atkinson, D. E., 1977, Adenine nucleotide concentrations and turnover rates, their correlation with biological activity in bacteria and yeast, *Adv. Microbiol. Physiol.* **15**:253–306.

Charyulu, P. B. B., Roamakrishna, C., and Rao, V. R., 1980, Effect of 2-aminobenzimidazole on nitrogen fixers from flooded soil and their nitrogenase activity, *Bull. Environ. Contam. Toxicol.* **25**:482–486.

Chatarpaul, L., Robinson, J. B., and Kaushik, N. K., 1979, Role of tubificid worms on nitrogen transformations in stream sediment, *J. Fish. Res. Board Can.* **36**:673–678.

Clesceri, L. S., Park, P. A., and Bloomfield, J. A., 1977, General model of microbial growth and decomposition in aquatic ecosystem, *Appl. Environ. Microbiol.* **33**:1047–1058.

Cleveland, M. E., 1983, Environmental factors affecting the sorption of pesticides to aquatic sediments, M.S. thesis, University of West Florida, Pensacola, 27 pp.

Cole, C. V., Elliott, E. T., Hunt, H. W., and Coleman, D. C., 1978, Trophic interactions in soils as they affect energy and nutrient dynamics. V. Phosphorus transformations, *Microb. Ecol.* 4:381–387.

Cole, L. K., Metcalf, R. L., and Sanborn, J. R., 1976, Environmental fate of insecticides in terrestrial model ecosystems, *Int. J. Environ. Stud.* 10:7–14.

Coleman, D. C., Cole, C. V., Hunt, H. W., and Klein, D. A., 1978a, Trophic interactions in soils as they affect energy and nutrient dynamics: Introduction, *Microb. Ecol.* 4:345–349.

Coleman, D. C., Anderson, R. V., Cole, C. V., Elliott, E. T., Woods, L., and Campion, M. K., 1978b, Trophic interactions in soils as they affect energy and nutrient dynamics. IV. Flows of metabolic and biomass carbon, *Microb. Ecol.* 4:373–380.

Colwell, P. R., 1978, Toxic effects of pollutants on microorganisms, in: *Principles of Ecotoxicology* (G. C. Butler, ed.), pp. 275–295, John Wiley, New York.

Connolly, J. P., and O'Connor, D. J., 1982, *WASTOX: Preliminary Estuary and Stream Version Documentation, Annual Report,* U.S. Environmental Protection Agency cooperative agreement CR807827-02, Gulf Breeze, Florida, 96 pp.

Cooke, G. D., 1967, The pattern of autotrophic succession in laboratory microcosms, *Bioscience* 17:717–721.

Cooke, G. D., 1974, Aquatic laboratory microsystems and communities, in: *Structure and Function of Fresh Water Microbial Communities* (J. Cairns, ed.), pp. 47–86, Virginia Polytechnic Institute, Blacksburg.

Cooper, W., Stout, J., and Boling, R., 1982, *Fate and Effects of p-Cresol in Outdoor Stream Channels, Annual Report,* U.S. Environmental Protection Agency cooperative agreement CR80755010, U.S. EPA Environmental Research Laboratory, Duluth, Minnesota, 86 pp.

Craib, J. S., 1965, A sampler for taking short undisturbed marine cores, *J. Cons. Cons. Perm. Int. Explor. Mer.* 30:34–39.

Cummins, K. W., 1974, Structure and function of stream ecosystems, *Bioscience* 24:-631–641.

Davies, J. M., and Gamble, J. C., 1979, Experiments with large enclosed ecosystems, *Philos. Trans. R. Soc. London Ser. B.* 286:523–574.

DePinto, J. V., Guminiak, R. F., Howell, R. S., and Edzwald, J. K., 1980, Use of microcosms to evaluate acid lake recovery techniques, in: *Microcosms in Ecological Research* (J. P. Giesy, ed.), Department of Energy Symposium Ser. No. 52 (conf-781101), pp. 562–582, NTIS, Springfield, Virginia.

Draggan, S., 1977, Effects of substrate type and arsenic dosage level on arsenic behavior in grassland microcosms. I. Preliminary results on [74]As transport, in: *Terrestrial Microcosms and Environmental Chemistry* (J. M. Witt and J. W. Gillet, eds.), pp. 102–110, National Science Foundation, NSF/RA 79-0026.

Draggan, S., 1979, The role of microcosms in ecological research, *Int. J. Environ. Stud.* 13:83–182.

Dudzik, M., Harte, J., Jassby, A., Lapan, E., Levy, D., and Rees, J., 1979, Some considerations in the design of aquatic microcosms for plankton studies, *Int. J. Environ. Stud.* 13:125–130.

Edberg, N., and Hofsten, B., 1973, Oxygen uptake of bottom sediments studied *in situ* and in the laboratory, *Water Res.* 7:1285–1294.

Edwards, R. W., and Rolley, H. L. J., 1965, Oxygen consumption of river muds, *J. Ecol.* 53:1–19.

Elliott, E. T., Cole, C. V., Coleman, D. C., Anderson, R. V., Hunt, H. W., and McClellan, J. F., 1979, Amoebal growth in soil microcosms, a model system of C_1 N and P trophic dynamics, *Int. J. Environ. Stud.* 13:169–174.

Elmgren, R., Vargo, G. A., Grassle, J. F., Grassle, J. P., Heinle, D. R., Langlois, G., and Vargo, S. L., 1980, Trophic interactions in experimental marine ecosystem perturbed by oil, in:

Microcosms in Ecological Research (J. P. Giesy, ed.), pp. 779–800, U.S. Department of Energy Symposium Ser. 52 (CONF-781101), NTIS, Springfield, Virginia.

Elner, J. K., Wildish, D. J., and Johnston, D. W., 1981, Fate of sprayed formulated aminocarb in freshwater, *Chemosphere* **10**:1025–1034.

Falco, J. W., Sampson, K. T., and Carsel, R. F., 1977, Physical modeling of pesticide degradation, *Dev. Ind. Microbiol.* **18**:193–202.

Fenchel, T. M. and Harrison, P., 1976, The significance of bacterial grazing and mineral cycling for the decomposition of particulate detritus, in: *The Role of Terrestrial and Aquatic Organisms in Decomposition Processes* (J. M. Anderson, ed.), pp. 285–299, Blackwell, Oxford, England.

Flint, R. W., and Goldman, C. R., 1975, The effects of a benthic grazer on the primary productivity of the littoral zone of Lake Tahoe, *Limnol. Oceanogr.* **20**:935–944.

Flint, R. W., Duke, T. W., and Kalke, R. D., 1978, Benthos investigations: Sediment boxes or natural bottom, *Bull. Environ. Contam. Toxicol.* **28**:257–265.

Focht, D. D., and Verstraete, W., 1977, Biochemical ecology of nitrification and denitrification, in: *Advances in Microbial Ecology*, Vol. 1 (M. Alexander, ed.), pp. 135–214, Plenum Press, New York.

Fry, J. C., 1982, Interaction between bacteria and benthic invertebrates, in: *Sediment Microbiology* (D. B. Nedwell and C. M. Brown, eds.), pp. 171–201, Academic Press, New York.

Fry, J. C., and Ramsey, A. J., 1977, Changes in the activity of epiphytic bacteria of *Elodea canadensis* and *Char vulgaris* flowing treatment with the herbicide paraquat, *Limnol. Oceanogr.* **22**:556–561.

Gee, J. H., and Bartnik, V. G., 1969, Simple stream tank simulating a rapids environment, *J. Fish. Res. Board Can.* **26**:2227–2230.

Gerike, P., and Fischer, W. K., 1979, A correlation study of the biodegradability determinations with various chemicals in various tests, *Ecotoxicol. Environ. Safety* **3**:159–173.

Giddings, J. M., and Eddleman, G. K., 1977, The effects of microcosm size and substrate type on aquatic microcosm behavior and arsenic transport, *Arch. Environ. Contam. Toxicol.* **6**:491–505.

Giddings, J. M., and Eddleman, G. K., 1978, Photosynthesis/respiration ratios in aquatic microcosms under arsenic stress, *Water Air Soil Pollut.* **9**:207–212.

Giddings, J. M., Walton, B. T., Eddleman, G. K., and Olson, K. G., 1979, Transport and fate of anthracene in aquatic microcosms, in: *Workshop: Microbial Degradation of Pollutants in Marine Environments* (A. W. Bourquin and P. H. Pritchard, eds.), pp. 312–320, U.S. Environmental Protection Agency, EPA-600/9-79-012.

Giesy, J. P., 1978, Cadmium inhibition of leaf decomposition in an aquatic microcosm, *Chemosphere* **6**:467–475.

Giesy, J. P. (ed.), 1980, *Microcosms in Ecological Research,* Symposium, Savannah River Ecology Laboratory, Augusta, Georgia, U.S. Department of Energy Symposium Ser. 52 (Conf-781101), NTIS, Springfield, Virginia, 1110 pp.

Gillett, J. W., and Gile, J. D., 1976, Pesticide fate in terrestrial laboratory ecosystems, *Int. J. Environ. Stud.* **10**:15–22.

Goodyear, C. P., Boyd, C. E., and Beyers, R. J., 1972, Relationships between primary productivity and mosquitofish *(Gambusia affinis)* production in large microcosms, *Limnol. Oceanogr.* **17**:445–450.

Goulder, R., Blanchard, A. S., Sanderson, P. L., and Wright, B., 1978, A note on the recognition of pollution stress in populations of estuarine bacteria, *J. Appl. Bacteriol.* **46**:285–289.

Goulder, R., Blanchard, A. S., Sanderson, P. L., and Wright, B., 1980, Relationships between heterotrophic bacteria and pollution in an industrial estuary, *Water Res.* **14**:591–601.

Graetz, D. A., Chesters, G., Daniel, T. C., Newland, L. W., and Lee, G. B., 1970, Parathion degradation in lake sediments, *J. Water Pollut. Control Fed.* **2**:R76–R94.

Graetz, D. A., Keeney, D. R., and Aspiras, R. B., 1973, Eh status of lake sediment-water systems in relation to nitrogen transformations, *Limnol. Oceanogr.* **18**:908–1017.

Greaves, M. P., Davies, H. A., Marsh, J. A. P., and Wingfield, G. I., 1976, Herbicides and soil microorganisms, *Crit. Rev. Microbiol.* **5**:1–38.

Griffiths, R. P., Caldwell, B. A., Broich, W. A., and Morita, R. Y., 1982, Long-term effects of crude oil on microbial processes in subarctic marine sediments: Studies on sediments amended with organic nutrients, *Mar. Pollut. Bull.* **13**:273–278.

Hammonds, A. S., 1981, *Methods for Ecological Toxicology: A Critical Review of Laboratory Multispecies Tests,* U.S. Environmental Protection Agency, EPA-56-/11-80-026, 307 pp.

Hansen, J. I., Henriksen, K., and Blackburn, T. H., 1981, Seasonal distribution of nitrifying bacteria and rates of nitrification in coastal marine sediments, *Microb. Ecol.* **7**:297–304.

Hansen, S. R., and Garton, R. R., 1982, The effects of diflubenzuron on a complex laboratory stream community, *Arch. Environ. Contam. Toxicol.* **11**:1–10.

Hardy, J. T., and Valett, M., 1981, Natural and microcosm phytoneuston communities of Sequin Bay, Washington, *Estuarine Coastal Shelf Sci.* **12**:3–12.

Hargrave, B. T., 1970, The effect of a deposit-feeding amphipod on the metabolism of benthic microflora, *Limnol. Oceanogr.* **15**:21–30.

Hargrave, B. T., 1976, The central role of invertebrate forces in sediment decomposition, in: *Role of Terrestrial and Aquatic Organisms in Decomposition Processes* (J. M. Anderson and A. Macfadyen, eds.), pp. 301–321, Blackwell, Oxford, England.

Harris, W. F., 1980, *Microcosms as Potential Screening Tools for Evaluating Transport and Effects of Toxic Substances,* U.S. Environmental Protection Agency, EPA-600/3-80-092, 379 pp.

Harrison, P. G., 1977, Decomposition of macrophyte detritus in seawater: Effects of grazing by amphipods, *Oikos* **28**:165–169.

Harrison, P. G., and Mann, K. H., 1975, Detritus formation from eelgrass *(Zostera marina):* The relative effects of fragmentation, leaching and decay, *Limnol. Oceanogr.* **20**:924–934.

Harte, J., Levy, D., Rees, J., and Saegebarth, E., 1980, Making microcosms an effective assessment tool, in: *Microcosms in Ecological Research* (J. P. Giesy, ed.), pp. 105–137, U.S. Department of Energy Symposium Ser. 52 (Conf-781101), NTIS, Springfield, Virginia.

Harty, B., and McLachlan, A., 1982, Effects of water-soluble fractions of crude oil and dispersants on nitrate generation by sandy beach microflora, *Mar. Pollut. Bull.* **13**:287–291.

Hauxhurst, J. D., Kaneko, T., and Atlas, R. M., 1981, Characteristics of bacterial communities in the Gulf of Alaska, *Microb. Ecol.* **7**:167–182.

Heath, R. T., 1979, Holistic study of an aquatic microcosm: Theoretical and practical implications, *Int. J. Environ. Stud.* **13**:87–93.

Henriksen, K., 1980, Measurement of *in situ* rates of nitrification in sediment, *Microb. Ecol.* **6**:329–337.

Henriksen, K., Hansen, J. I., and Blackburn, T. H., 1980, The influence of benthic infauna on exchange rates of inorganic nitrogen between sediment and water, *Ophelia Suppl.* **1**:249–256.

Henriksen, K., Hansen, J. I., and Blackburn, T. H., 1981, Rates of nitrification, distribution of nitrifying bacteria and nitrate fluxes in different types of sediment from Danish waters, *Mar. Biol.* **61**:299–304.

Herzberg, M. A., Klein, A., and Coleman, D. C., 1978, Trophic interactions in soils as they affect energy and nutrient dynamics. II. Physiological responses of selected rhizosphere bacteria, *Microb. Ecol.* **4**:351–359.

Hill, J., 1979, Mathematical modeling of pesticides in the environment: Current and future developments, *J. Environ. Systems* **9**:99–107.

Hill, J., and Wiegert, R. G., 1980, Microcosms in ecological modeling, in: *Microcosms in Ecological Research* (J. P. Giesy, ed.), pp. 138–163, U.S. Department of Energy Symposium Ser. 52 (Conf-781101), NTIS, Springfield, Virginia.

Hopkinson, C. S., and Day, J. W., 1977, A model of the Barataria Bay salt marsh ecosystem, in: *Ecosystem Modeling in Theory and Practice* (A. S. Hall and J. W. Day, eds.), pp. 236–265, Wiley-Interscience, New York.

Howard, P. H., Saxena, J., and Sikka, H., 1978, Determining the fate of chemicals, *Environ. Sci. Technol.* **12**:398–407.

Howes, B. C., Howarth, R. W., Teal, J. M., and Valiela, I., 1981, Oxidation–reduction potentials in a salt marsh: Spatial patterns and interactions with primary production, *Limnol. Oceanogr.* **26**:350–360.

Hsu, T. S., and Bartha, R., 1979, Accelerated mineralization of two organophosphate insecticides in the rhizosphere, *Appl. Environ. Microbiol.* **37**:36–41.

Hylleberg, J., and Henriksen, K., 1980, The central role of bioturbation in sediment mineralization and element cycling, *Ophelia Suppl.* **1**:1–16.

Isensee, A. R., 1976, Variability of aquatic model ecosystem-derived data, *Int. J. Environ. Stud.* **10**:35–41.

Isensee, A. R., Kearney, P. C., Woolson, E. A., Jones, G. E., and Williams, V. P., 1973, Distribution of alkyl arsenicals in the model ecosystem, *Environ. Sci. Technol.* **7**:841–845.

Jackson, D. R., Washburne, C. D., and Ausmus, B. S., 1977, Loss of Ca and NO_3-N from terrestrial microcosms as an indicator of soil pollution, *Water Air Soil Pollut.* **8**:279–284.

Jackson, D. R., Ausmus, B. S., and Levin, M., 1979, Effects of arsenic on nutrient dynamics of grassland microcosms and field plots, *Water Air Soil Pollut.* **11**:13–21.

Jahnke, R. A., Emerson, S. R., and Murray, J. W., 1982, A model of oxygen reduction, denitrification, and organic matter mineralization in marine sediments, *Limnol. Oceanogr.* **27**:610–623.

Jassby, A., Dudzik, M., Rees, J., Lapan, E., Levy, D., and Harte, J., 1977a, *Production Cycles in Aquatic Microcosms*, U.S. Environmental Protection Agency, EPA-600/7-77-077, 51 pp.

Jassby, A., Rees, J., Dudzik, M., Levy, D., Lapan, E., and Harte, J., 1977b, *Trophic Structure Modifications by Planktovorous Fish in Aquatic Microcosms*, U.S. Environmental Protection Agency, EPA-600/7-77-096, 18 pp.

Jensen, S., and Jernelov, A., 1969, Biological methylation of mercury in aquatic organisms, *Nature (London)* **223**:753–754.

Jernelov, A., 1978, Release of methyl mercury from sediments with layers containing inorganic mercury at different depths, *Limnol. Oceanogr.* **15**:958–960.

Jones, J. G., 1979, Microbial nitrate reduction in freshwater sediments, *J. Gen. Microbiol.* **115**:27–35.

Jones, J. G., 1982, Activities of aerobic and anaerobic bacteria in lake sediments and their effect on the water column, in: *Sediment Microbiology* (D. B. Nedwell and C. M. Brown, eds.), pp. 107–145, Academic Press, New York.

Jones, R. D., and Hood, N. A., 1980, The effects of organophosphorus pesticides on estuarine ammonia oxidizers, *J. Can. Microbiol.* **26**:1296–1299.

Juengst, F. W., and Alexander, M., 1975, Effect of environmental conditions on the degradation of DDT in model marine ecosystems, *Mar. Biol.* **33**:1–6.

Karickhoff, S. W., Brown, D. S., and Scott, T. A., 1979, Sorption of hydrophobic pollutants on natural sediments, *Water Res.* **13**:241–247.

Katan, J., Fuhremann, T. W., and Lichtenstein, E. P., 1976, Binding of [^{14}C]Parathion in soil: A reassessment of pesticide persistence, *Science* **193**:891–894.

Kloskowski, R., Schevnert, I., Klein, W., and Forte, F., 1981, Laboratory screening of distribution, conversion and mineralization of chemicals in the soil-plant-system and comparison to outdoor experimental data, *Chemosphere* **10**:1089–1100.

Kremer, J. N., 1979, An analysis of the stability characteristics of an estuarine ecosystem model, in: *Marsh–Estuarine Systems Simulations* (R. F. Dame, ed.), pp. 189–206, University of South Carolina Press, Columbia.

Larsson, V., and Hastrom, A., 1979, Phytoplankton exudate release as an energy source for the growth of pelagic bacteria, *Mar. Biol.* **52**:199–206.

Lassiter, R. R., 1975, *Modeling the Dynamics of Biological and Chemical Components of Aquatic Ecosystems*, U.S. Environmental Protection Agency, EPA-660/3-75-012, 54 pp.

Lassiter, R. R., 1979, Microcosms as ecosystem for testing ecological models, in: *State-of-the-Art in Ecological Modeling*, Vol. 7 (S. E. Jorgensen, ed.), pp. 127–161, Pergamon Press, Oxford.

Lassiter, R. R., Baughman, G. L., and Burns, L. A., 1978, Fate of toxic organic substances in the aquatic environment, in: *State-of-the-Art in Ecological Modeling*, Vol. 7 (S. E. Jorgensen, ed.), pp. 219–295, Pergamon Press, Oxford.

Lauff, G. H., and Cummins, K. W., 1964, A model stream for studies in lotic ecology, *Ecology* **45**:188–191.

Lee, R. F., and Ryan, C., 1979, Microbial degradation of organochlorine compounds in estuarine waters and sediments, in: *Microbial Degradation of Pollutants in Marine Environments* (A. W. Bourquin and P. H. Pritchard, eds.), pp. 443–450, U.S. Environmental Protection Agency, EPA-600/9-79-012.

Lee, R. F., Gardner, W. S., Anderson, J. W., Blaylock, J. W., and Barwell-Clarke, J., 1978, Fate of polycyclic aromatic hydrocarbons in controlled ecosystem enclosures, *Environ. Sci. Technol.* **12**:832–838.

Levandowsky, M., 1977, Multispecies cultures and microcosms, in: *Marine Ecology*, Vol. III (O. Kinne, ed.), pp. 1399–1452, John Wiley, New York.

Levin, S. A., 1982, *New prospectives in Ecotoxicology, Workshop Report*, Ecosystems Research Center, Cornell University, Ithaca, New York, 125 pp.

Lewis, D. L., and Holm, H. W., 1981, Rates of transformation of methyl parathion and diethyl phthalate by aufwuchs microorganisms, *Appl. Environ. Microbiol.* **42**:698–703.

Lichtenstein, E. P., Liang, T. T., and Fuhremann, T. W., 1978, A compartmentalized microcosm for studying the fate of chemicals in the environment, *J. Agric. Food, Chem.* **26**:948–953.

Liu, D., Thomson, K., and Stachan, W. M., 1980, Biodegradation of carbaryl in simulated aquatic environments, *Bull. Environ. Contam. Toxicol.* **27**:412–417.

Liu, D., Thomson, K., and Strachan, W. M., 1981, Biodegradation of pentachlorophenol in a simulated aquatic environment, *Bull. Environ. Contam. Toxicol.* **26**:85–90.

Lopez, G. R., Levinton, J. S., and Slobodkin, L. B., 1977, The effect of grazing by the detritivore *Orchestia grillus* on *Spartina* litter and its associated microbial community, *Oecologia (Berlin)* **30**:111–127.

Maki, A. W., 1980, Evaluation of toxicant effects on structure and function of model stream communities: Correlation with natural effects, in: *Microcosms in Ecological Research* (J. P. Giesy, ed.), pp. 583–609, Department of Energy Symposium Ser. 52 (Conf-781101), NTIS.

Mann, K. H., 1979, Qualitative aspects of estuarine modeling, in: *Marsh–Estuarine Systems Simulation* (R. F. Dame, ed.), pp. 207–220, University of South Carolina Press, Columbia.

Marshall, W. K., and Roberts, J. R., 1971, Simulation modeling of the distribution of pesticides in ponds, Nat. Res. Counc. Can. NRCC/CNRR 16073 **2**:253–278.

Martin, Y. P., and Bianchi, M. A., 1980, Structure, diversity and catabolic potentialities of aerobic heterotrophic bacterial populations associated with continuous cultures of natural marine phytoplankton, *Microb. Ecol.* **5**:265–279.

McCall, P. L., and Fisher, J. B., 1980, Effects of tubificid oligochaetes on physical and chemical properties of Lake Erie sediments, in: *Aquatic Oligochaete Biology* (R. O. Brinkhurst and D. G. Cook, eds.), pp. 253–317, Plenum Press, New York.

McIntire, C. D., 1964, Primary production in laboratory streams, *Limnol. Oceanogr.* **9**:92–102.

McIntire, C. D., 1965, Structural characteristics of benthic algal communities in laboratory streams, *Limnol. Oceanogr.* **9**:92–102.

McIntire, C. D., 1966, Some factors affecting respiration of periphyton communities in lotic environments, *Ecology* **47**:918–930.

McIntire, C. D., 1978, Periphyton assemblages in laboratory streams, in: *River Ecology* (B. A. Whitton, ed.), pp. 403–430, University of California Press, Berkeley.

McIntire, C. D., Colby, J. A., and Hall, J. D., 1975, The dynamics of small lotic ecosystems: A modeling approach, *Verh. Int. Verein. Limnol.* **19**:1599–1609.

McKinley, K. R., and Wetzel, R. G., 1979, Photolithotrophy, photoheterotrophy and chemoheterotrophy: Patterns of resource utilization on an annual and a diurnal basis within pelagic microbial communities, *Microb. Ecol.* **5**:1–15.

Mehran, M., and Tanji, K. K., 1974, Computer modeling of nitrogen transformations in soil, *J. Environ. Qual.* **3**:391–410.

Metcalf, R. L., Sangha, G. K., and Kapoor, I. P., 1971, Model ecosystem for the evaluation of pesticide biodegradability and ecological magnification, *Environ. Sci. Technol.* **5**:709–713.

Metcalf, L., Kapoor, P., Schuth, C. K., and Sherman, P., 1973, Model ecosystem studies of the environmental fate of six organochlorine pesticides, *Environ. Health Perspect.* **4**:35–44.

Nash, R. G., and Beall, M. L., 1977, A microagroecosystem to monitor the environmental fate of pesticides, in: *Terresterial Microcosms and Environmental Chemistry* (J. M. Witt and J. W. Gillett, eds.), pp. 86–94, National Science Foundation, NSF/RA 79-0026.

Nash, R. G., Beall, M. L., and Harris, W. G., 1977, Toxaphene and 1,1,1,-trichloro -2,2-bis (*p*-chlorophenyl) ethane (DDT) losses from cotton in an agroecosystem chamber, *J. Agric. Food Chem.* **25**:336–341.

Nelson, J. D., and Colwell, R. R., 1975, The ecology of mercury-resistant bacteria in Chesapeake Bay, *Microb. Ecol.* **2**:191–218.

Nixon, S. W., 1981, Remineralization and nutrient cycling in coastal marine ecosystems, in: *Estuaries and Nutrients* (B. J. Neilson and L. E. Cronin, eds.), pp. 111–138, Humana Press, Clifton, New Jersey.

Nixon, S. W., and Kremer, J. N., 1977, Narragansett Bay—The development of a composite simulation model for a New England estuary, in: *Ecosystem Modeling in Theory and Practice* (C. A. S. Hall and J. W. Day, eds.), pp. 622–673, John Wiley, New York.

Nixon, S. W., Oviatt, C. A., Kremer, J. N., and Perez, K., 1979, The use of numerical models and laboratory microcosms in estuarine ecosystem analysis—simulations of a winter phytoplankton bloom, in: *Marsh–Estuarine Systems Simulations* (R. F. Dame, ed.), pp. 165–188, University of South Carolina Press, Columbia.

Odum, E. P., 1969, The strategy of ecosystem development, *Science* **164**:262–270.

Olanczuk-Neyman, K. M., and Vosjan, J. H., 1977, Measuring respiratory electron-transport-system activity in marine sediments, *Neth. J. Sea Res.* **11**:1–13.

Olsen, B. H., and Cooper, R. C., 1976, Comparison of aerobic and anaerobic methylation of mercury chloride by San Fransico Bay sediments, *Water Res.* **10**:113–116.

O'Neill, R. B., Ausmus, B. S. Jackson, D. R., Van Hook, R. I., Van Voris, P., Washburne, C., and Watson, A. P., 1977, Monitoring terrestrial ecosystems by analysis of nutrient export, *Water Air Soil Pollut.* **8**:271–277.

Orndorff, S. A., and Colwell, R. R., 1980, Effect of Kepone on estuarine microbial activity, *Microb. Ecol.* **6**:357–368.

Painter, H. A., 1970, A review of literature of inorganic nitrogen metabolism in microorganisms, *Water Res.* **4**:393.

Pamatmat, M. M., 1971, Oxygen consumption by the seabed. IV. Shipboard and laboratory experiments, *Limnol. Oceanogr.* **16**:536–550.

Pamatmat, M. M., and Bhagwat, A. M., 1973, Anaerobic metabolism in Lake Washington sediments, *Limnol. Oceanogr.* **18**:611–627.

Paris, D. F., Steen, W. C., Baughman, G. L., and Barnett, J. T., 1981, Second-order model to predict microbial degradation of organic compounds in natural waters, *Appl. Environ. Microbiol.* **41**:603–609.

Patten, B. C., and Witkamp, M., 1967, Systems analysis of [134]cesium kinetics in terrestrial microcosms, *Ecology* **48**:813–824.

Perez, K. T., Morrison, G. M., Lackie, N. F., Oviatt, C. A., Nixon, S. W., Buckley, B. A., and Heltshe, J. F., 1977, The importance of physical and biotic scaling to the experimental simulation of a coastal marine ecosystem, *Helgol. Wis. Meeresunters.* **30**:144–162.

Peterson, R. C., and Cummins, K. W., 1974, Leaf processing in a woodland stream, *Fresh Water Biol.* **4**:343–368.

Pfaender, F. K., and Alexander, M., 1972, Extensive microbial degradation of DDT *in vitro* and DDT metabolism by natural communities, *Agric. Food Chem.* **20**:842–846.

Pilson, M. E. Q., Oviatt, C. A., Vargo, G. A., and Vargo, S. L., 1979, Replicability of MERL microcosms: Initial observations, in: *Advances in Marine Environmental Research* (F. S. Jacoff, ed.), U.S. Environmental Protection Agency, EPA-600/9-79-035, 409 pp.

Pilson, M. E. Q., Oviatt, C. A., and Nixon, S. W., 1980, Annual nutrient cycles in a marine microcosm, in: *Microcosms in Ecological Research* (J. P. Giesy, ed.), pp. 753–778, U.S. Department of Energy Symposium Ser. 52 (Conf-7811-1), NTIS, Springfield, Virginia.

Portier, R. J., and Meyers, S. P., 1981, Chitin transformation and pesticide interactions in a simulated aquatic microenvironmental system, *Dev. Ind. Microbiol.* **22**:543–555.

Pritchard, P. H., 1981, Model ecosystems, in: *Environmental Risk Analysis for Chemicals* (R. A. Conway, ed.), pp. 257–353, Van Nostrand Reinhold, New York.

Pritchard, P. H., and Cripe, C. R., 1983, A microcosm system to model the fate and effects of *p*-cresol and other pollutants in lotic stream ecosystems, *Limnol. Oceanogr.* (submitted).

Pritchard, P. H., and Van Veld, P., 1983, Evidence for biodegradation of *p*-cresol in outdoor stream channels, *J. Soc. Environ. Toxicol. Chem.* (submitted).

Pritchard, P. H., Bourquin, A. W., Frederickson, H. L., and Maziarz, T., 1979, System design factors affecting environmental fate studies in microcosms, in: *Microbial Degradation of Pollutants in Marine Environments* (A. W. Bourquin and P. H. Pritchard, eds.), pp. 251–272, U.S. Environmental Protection Agency, EPA-600/9-79-012.

Pritchard, P. H., Van Veld, P., and Boyer, J. M., 1983a, Comparisons of the rate of *p*-cresol degradation in shake flasks, microcosms and field streams, *Appl. Environ. Microbiol.* (submitted).

Pritchard, P. H., Connolly, J. P., Maziarz, T. M., and Bourquin, A. W., 1983b, Application of microcosm studies to verify chemical fate assessments: Comparison of the fate of methyl parathion in a sediment–water system, *Water Res.* (in press).

Rubinstein, N. I., 1979, A benthic bioassay using time-lapse photography to measure the effect of toxicants on the feeding behavior of lugworms (Polychaeta: Arenicolidae), in: *Marine Pollution: Functional Responses* (W. B. Vernberg, A. Calabrese, F. Thurberg, and F. J. Vernberg, eds.), pp. 341–351, Academic Press, New York.

Salt, G. W., 1979, A comment on the use of the term *emergent properties, Am. Nat.* **113**:145–148.

Sayler, G. S., Lund, L. C., Shiaris, M. P., Sherrill, T. W., and Perkins, R. E., 1979, Comparative effects of aroclor 1254 (polychorinated biphenyls) and phenanthrene on glucose uptake by freshwater microbial populations, *Appl. Environ. Microbiol.* **37**:878–885.

Schindler, J. E., Waide, J. B., Waldron, M. C., Hains, J. J., Schreiner, S. P., Freedman, M. L., Benz, S. L., Pattigrew, D. R., Schissel, L. A., and Clark, P. J., 1980, A microcosm approach to the study of biogeochemical systems. 1. Theoretical rationale, in: *Microcosms in Ecological Research* (J. P. Giesy, ed.), pp. 192–203, U.S. Department of Energy Symposium Ser. 52 (Conf-781101), NTIS, Springfield, Virginia.

Seitzinger, S., Nixon, S., Pilson, M., and Burke, S., 1980, Denitrification and nitrous oxide production in near-shore marine sediments, *Geochim. Cosmochim. Acta* **44**:1853–1860.

Sethunathan, N., Siddaramapa, R., Rajaram, K. P., Barik, S., and Wahid, P. A., 1977, Parathion: Residues in soil and water, *Residue Rev.* **68**:91–122.

Shaw, B., and Hopke, P. K., 1975, The dynamics of diaquat in a model eco-system, *Environ. Lett.* **8**:325–335.

Sikka, H. C., and Rice, C. P., 1973, Persistence of endothall in aquatic environment as determined by gas–liquid chromatography, *J. Agric. Food Chem.* **21**:842–846.

Simsiman, G. V., and Chesters, G., 1976, Persistence of diquat in the aquatic environment, *Water Res.* **10**:105–112.

Smith, G. A., Nickels, J. S., Bobbie, R. J., Richards, N. L., and White, D. C., 1982, Effects of oil and gas well-drilling fluids on the biomass and community structure of microbiota that colonize sands in running seawater, *Arch. Environ. Contam. Toxicol.* **11**:17–23.

Spain, J. C., Pritchard, P. H., and Bourquin, A. W., 1980, Effects of adaptation on biodegradation rates in sediment/water cores from estuarine and freshwater environments, *Appl. Environ. Microbiol.* **40**:726–734.

Stay, F. S., 1980, *Review of Aquatic Microcosms Techniques Used for Hazard Assessment of Potentially Toxic Compounds,* U.S. Environmental Protection Agency, Environmental Research Laboratory, Corvallis, Oregon, Publication 052, 33 pp.

Steel, J. H., and Menzel, D. W., 1978, The application of plastic enclosures to the study of pelagic marine biota, *Rapp. P.V. Reun. Cons. Int. Explor. Mer.* **173**:7–12.

Stout, J. D., 1980, The role of protozoa in nutrient cycling and energy flow, in: *Advances in Microbial Ecology,* Vol. 4 (M. Alexander, ed.), pp. 1–50, Plenum Press, New York.

Straskrabova, V., and Fuksa, J., 1982, Diel changes in numbers and activities of bacterioplankton in a reservoir in relation to algal production, *Limnol. Oceanogr.* **27**:660–672.

Tempest, D. W., 1970, The place of continuous culture in microbial research, *Adv. Microbiol. Physiol.* **4**:223–250.

Titus, J. A., Parsons, J. E., and Pfister, R. M., 1980, Translocation of mercury and microbial adaptation in a model aquatic system, *Bull. Environ. Contam. Toxicol.* **25**:456–464.

Troussellier, M., and Legendre, P., 1981, A functional evenness index for microbial ecology, *Microb. Ecol.* **7**:283–296.

Tsushimoto, G., Matsumura, F., and Sago, R., 1982, Fate of 2,3,7,8-tetrachlorodibenzo-*p*-dioxin (TCDD) in an outdoor pond and in a model aquatic ecosystem, *Environ. Toxicol. Chem.* **1**:61–68.

Tu, C. M., 1980, Influence of pesticides and some of the oxidized analogues on microbial populations, nitrification and respiration activities in soil, *Bull. Environ. Contam. Toxicol.* **24**:13–19.

Tu, C. M., and Miles, J. R. W., 1976, Interactions between insecticides and soil microbes, *Residue Rev.* **64**:17–65.

Twinch, A. J., and Breen, C. M., 1981, The study of phosphorus and nitrogen fluxes in enriched isolation columns, *Hydrobiologia* **77**:49–60.

Vanderborght, J. P., and Billen, G., 1975, Vertical distribution of nitrate concentration in interstitial water of marine sediments with nitrification and denitrification, *Limnol. Oceanogr.* **20**:953–961.

Van Voris, P., O'Neill, R. V., Emanuel, W. R., and Shugart, H. H., 1980, Function complexity and ecosystem stability, *Ecology* **61**:1352–1360.

Virtanen, M. T., Kihlstrom, M., Roos, A., and Kainulainen, H., 1982, Model ecosystem for environmental transport of xenobiotics, *Arch. Environ. Contam. Toxicol.* **11**:410–424.

Vosjan, J. H., and Olanczuk-Neyman, K. M., 1977, Vertical distribution of mineralization processes in a tidal sediment, *Neth. J. Sea Res.* **11**:14–23.

Waide, J. B., Schindler, J. E., Waldron, M. C., Hains, J. J., Schreiner, S. P., Freedman, M. L., Benz, S. L., Pettigrew, D. R., Schissel, L. A., and Clark, J. P., 1980, A microcosm approach to the study of biogeochemical systems: Responses of aquatic laboratory microcosms to physical, chemical and biological perterbations, in: *Microcosms in Ecological Research* (J. B. Giesy, ed.), pp. 204–223, U.S. Department of Energy Symposium Ser. 52 (Conf-781101), NTIS, Springfield, Virginia.

Wangersky, P. J., 1978, Production of dissolved organic matter, in: *Marine Ecology*, Vol. IV (O. Kinne, ed.), pp. 115–220, John Wiley, New York.

Warren, C. E., and Davis, G. E., 1971, Laboratory stream research: Objectives, possibilities and constraints, *Annu. Rev. Ecol. Syst.* **2**:111–144.

Webb, J. E., and Theodor, J. L., 1972, Wave induced circulation in submerged sands, *J. Mar. Biol. Assoc. U.K.* **52**:903–914.

Weiss, P. A., 1971, The basic concept of hierarchic systems, in: *Hierarchically Organized Systems in Theory and Practice* (P. A. Weiss, ed.), pp. 1–43, Hafner, New York.

White, D. C., Bobbie, R. J., King, J. D., Nickels, J. S., and Amoe, P., 1979, Lipid analysis of the sediments for microbial biomass and community structure, in: *Methodology for Biomass Determinations and Microbial Activities in Sediments* (C. D. Litchfield and P. L. Seyfried, eds.), pp. 87–103, American Society for Testing and Materials, Philadelphia, Pennsylvania.

Widdus, R., Trudgill, P. W., and Turnell, D. C., 1971, Effects of technical chlordane on growth and energy metabolism of *Streptococcus faecalis* and *Mycobacterium phlei:* A comparison with *Bacillus subtilis, J. Gen. Microbiol.* **69**:21–23.

Wiebe, W. J., and Smith, D. F., 1977, Direct measurement of dissolved organic carbon release by plankton and incorporation by microheterotrophs, *Mar. Biol.* **42**:213–223.

Wiegert, R. G., Christian, R. R., Gallagher, J. L., Hall, J. R., Jones, R. D., and Wetzel, R. L., 1975, A preliminary ecosystem model of a coastal Georgia *Spartina* marsh, *Estuarine Res.* **1**:583–601.

Witherspoon, J. P., Bondietti, E. A., Draggon, S., Taub, F., Pearson, P., and Trabokla, J. R., 1976, *State-of-the-Art and Proposed Testing for Environmental Transport of Toxic Substances*, U.S. Environmental Protection Agency, EPA-500/5-76-001, 105 pp.

Witkamp, M., 1976, Microcosm experiments on element transfer, *Int. J. Environ. Stud.* **10**:59–63.

Witkamp, M., and Ausmus, B. S., 1975, Processes in decomposition and nutrient transfer in forest systems, in: *The Role of Terrestrial and Aquatic Organisms in Decomposition Processes* (J. M. Anderson and A. Macfadyen, eds.), pp. 397–416, Blackwell, Oxford, England.

Witt, J. M., and Gillett, J. W., 1977, *Terrestrial Microcosms and Environmental Chemistry, Proceedings of Symposium,* Corvallis, Oregon, National Science Foundation, NSF/RA 79-0026, 147 pp.

Wolfe, N. L., Zepp, R. G., Gordon, J. A., Baughman, G. L., and Cline, D. M., 1977, Kinetics of chronical degradation of malathion in water, *Environ. Sci. Tech.* **11**:88–93.

Wolfe, N. L., Zepp, R. G., Schlotzhaver, P., and Sink, M., 1982, Transformation pathways of hexachlorocyclopentadiene in the aquatic environment, *Chemosphere* **11**:91–101.

Yockim, R. S., Isensee, A. S., and Weber, E. A., 1980, Behavior of trifluralin in aquatic model ecosystems, *Bull. Environ. Contam. Toxicol.* **24**:134–141.

Index